Communication by Children and Adults

Sage's *Series in Interpersonal Communication* is designed to capture the breadth and depth of knowledge emanating from scientific examinations of face-to-face interaction. As such, the volumes in this series address the cognitive and overt behavior manifested by communicators as they pursue various conversational outcomes. The application of research findings to specific types of interpersonal relationships (e.g., marital, managerial) is also an important dimension of this series.

SAGE SERIES IN INTERPERSONAL COMMUNICATION

Mark L. Knapp, Series Editor

Communication by Children and Adults

Social Cognitive and Strategic Processes

Edited by

Howard E. Sypher
James L. Applegate

Sage Series in Interpersonal Communication

Volume 5

SAGE PUBLICATIONS
Beverly Hills   London   New Delhi

For information address:

SAGE Publications, Inc.
275 South Beverly Drive
Beverly Hills, California 90212

SAGE Publications India Pvt. Ltd.
C-236 Defence Colony
New Delhi 110 024, India

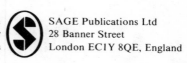

SAGE Publications Ltd
28 Banner Street
London ECIY 8QE, England

Printed in the United States of America

Library of Congress Cataloging in Publication Data

Main entry under title:

Communication by children and adults.

  (Sage series in interpersonal communication ; v. 5)
  1. Interpersonal communication.   2. Social perception.
3. Interpersonal communication in children.   4. Social
perception in children.   I. Sypher, Howard E.
II. Applegate, James L.   III. Series.
BF637.C45C643   1984   153.6   84-4803
ISBN 0-8039-2316-3
ISBN 0-8039-2315-5 (pbk.)

FIRST PRINTING

Contents

Series Editor's Foreword

For a long time we've known that it was important to understand human thought processes if we were ever to understand fully the nature of interpersonal communication. And there have always been a few courageous scholars who did not let the warnings of forbidding difficulty stand in the way of their exploring the workings of the "black box." But only in recent years have so many scholars from so many different academic disciplines tackled the questions pertaining to how we think, how thinking affects behavior, and how behavior affects thinking. The contributors to this volume provide readers with a broad and representative view of recent research and theory bearing on cognition and social interaction.

Each of the authors in this volume makes it clear that it is extremely difficult to conceptualize or do research in the area of interpersonal communication without having to face critical questions about the way communicators think—prior to, during, and after interaction. For example:

- To what extent are the communicators aware of their actions?
- To what extent can the communicators recall their interaction behaviors? How accurate are these recollections?
- How do the communicators form and organize their impressions of each other?
- How do the communicators form and organize their impressions of their own communication behavior?
- Out of all the stimuli available to communicators during interaction, which of them are attended to? Why?
- How do thoughts about social behavior influence the communicators' actual behavior? How do their behaviors influence their thoughts?

- How do communicators' cognitions of their partners, topic, self, and situation combine to affect the achievement of such goals as persuasion or comforting?

In short, virtually anything that has to do with the role of expectations, memory, awareness, fantasy, impression formation, information processing, information seeking, and knowledge structure as they relate to face-to-face interaction is likely to be found in this book. As extensive and central as the preceding list of processes appears, neither the editors nor the authors claim that an understanding of social cognition is anything more than a *part* of understanding the entire process of interpersonal communication. It is, however, clearly an important part.

The editors of this book have assembled an impressive group of selections. The frequent emphasis on the interface of actual communication behavior and cognitive processes is a feature that distinguishes this volume from many others on social cognition. The concern for actor cognitions as well as those of observers is another much-needed perspective for communication researchers. And the focus on cognition and communication of both children and adults in a single volume provides an opportunity for viewing these issues from a developmental perspective. *Communication by Children and Adults: Social Cognitive and Strategic Processes* is a welcome addition to our knowledge of cognition and communication and should be of considerable interest to psychologists (social, developmental, and cognitive) and communication scholars alike.

—*Mark L. Knapp*

Understanding Interpersonal Communication:
Integrating Research Trends

James L. Applegate and Howard E. Sypher

Interpersonal communication has rapidly developed into a diverse field of inquiry. Researchers are employing a variety of perspectives and methodological techniques. This volume focuses on one area of research concerned with analyzing the organization of and relationship between qualities of social cognition and communicative interaction. Researchers in psychology, communication, education, and linguisitics have all contributed to this developing area of study. Their work examines the quality of social cognition and communication during childhood, as well as individual differences among adults. Unfortunately, parallel developments in psychology and communication focusing on social cognitionn and interpersonal communication seem to be unrecognized. One purpose of this volume is to bring together some of these separate but similar research programs.

The chapters in the volume offer theoretical reviews defining conceptual issues of current concern and in many instances present original research addressing central issues in the area—for example, effects of social cognitive and communicative development on peer group relationships; prosocial and persuasive communication development; and relationship management in adult communication. The content is balanced between work done with children and adults. It should be of interest to students and researchers doing work in both areas. The volume also reflects the increasing concern in this area of research with schema concepts, particularly group and individual differences in schema or construct development and their effects on communication behavior. The authors' contributions reflect a developing awareness that communication behavior is not epiphenomenal to social cognition, but deserving of study in its own right—both as an influence on and as influenced by social cognitive processes.

The analyses of communication presented in this volume are framed around particular strategic functions of communication and the organization of interaction itself. The book reflects this functional framework by examining prosocial, comforting, and influence functions for children and relationship goals for adults. Individual contributions explore both the nature and social implications of differences in social cognitive and communicative behavior in various functional contexts.

The research in this volume also moves away from the use of general measures of social cognitive processes, such as global perspective-taking indices, and toward the use of more specific assessments of social cognitive structure. The latter measures are linked to equally specific and conceptually parallel developments in strategic behavior. This effort reflects a growing awareness of the multiple dimensions of cognitive and strategic development. It is also a response to the failure of research using global measures to evidence any consistent relationship between theoretically linked social cognitive and behavioral processes.

While there is considerable breadth in this volume, it also reflects a great deal of convergence in the conceptual and methodological issues addressed. This convergence is apparent even across child and adult studies.

In the area of research on social cognition and children's communication, Eisenberg and Silbereisen, as well as Barnett, provide well-reasoned accounts of conceptual and methodological problems plaguing research on empathy and perspective-taking as critical skills relevant to communication. Eisenberg and Silbereisen's survey of traditional approaches to the study of prosocial development emphasizes the limited success of this work in integrating the cognitive and behavioral components of prosocial behavior. They propose an "action theoretic" framework currently influencing European research which examines the multiple roles played by social cognition in influencing behavior. In elaborating their position, they suggest how we can better conceptualize the complex interplay between cognition and behavior through direct examination of the behavioral goals that inform action. That examination includes analyses of cultural, institutional, and developmental constraints on goal choice. The chapter concludes by integrating several research programs within the more comprehensive action theoretic approach.

Barnett gives particular attention to how the development of perspective-taking and empathic ability affects prosocial behavior. He argues that future research must integrate the cognitive (perspective-taking) and affective (empathic) components of social perception in studying the antecedents of prosocial behavior. Moreover, he contends that the development of these social perception skills does not constitute sufficient conditions for the production of prosocial behavior. Both Eisenberg and Silbereisen's and Barnett's suggestions for developing clear, differentiated conceptions of the social cognitive processes that are closely tied to independent but related features of communicative goals and action introduce a central theme that runs throughout the volume.

Burleson's chapter reviews a "constructivist" perspective on social cognition and communication grounded in Wernerian developmental theory. He shows the connection between social cognitive development and the acquisition of prosocial, comforting communication behaviors with distressed listeners. Burleson also assesses the strengths and weaknesses of the methods of data collection and analysis employed in current research. He provides a comprehensive review of the findings to date explaining differences in the development of comforting skills and the motivation to comfort. The chapter concludes with an integrative model to guide future research. In capturing the complex interplay between social cognitive development and the strategic competence to comfort, the model instantiates and extends many of the ideas forwarded in Eisenberg and Silbereisen's action theoretic framework. The research that Burleson reviews draws specific relations between social cognitive and communicative development in the prosocial domain and explores the mediating effects of contextual and motivational variables.

Forbes and Lubin offer an extension of the constructivist research on social cognition and persuasive communication development conducted by Jesse Delia and his associates. Their study relates the development of specific features of social reasoning to equally specific (and parallel) dimensions of strategic development. Based on the success of their analysis, Forbes and Lubin argue that future research must create very specific hierarchical coding systems that will capture parallel developments in social reasoning and the quality of strategies used to accomplish particular social goals. The authors' analysis of the relationship between children's reasoning and behavior, and their specific interpretation of results, should provide the interested reader in education or communication with considerable food for thought.

Oden, Wheeler, and Herzberger examine differences in the form and organization of conversations between children as they negotiate the use of a toy in a conflict-of-interest situation. Employing a speech act framework, they find interesting age-related differences in children's organization of conversations and in the types of strategies employed to negotiate conflict and to influence partners. The authors suggest that such developmental differences should be related to children's acceptance by peers. Their research goes beyond previous work in identifying the variety of influence strategies children employ in situated conversation and their probable social effects.

In another analysis of children's situated interactions, Rubin and Borwick provide an in-depth description of the types of request strategies children employ, as well as the types of responses particular strategies elicit. The authors directly assess the relation of differences in types of request strategies to children's relative sociability with peers. Their results suggest interesting differences in response-request patterns as a function of the general sociability of interactants.

In an extension of their well-known "communication game" approach, McCann and Higgins introduce the second half of the volume with an integrative review of research documenting the central role of communicative goals in organizing interpersonal interaction. They show that interactants' communicative goals systematically vary as a function of their socioeconomic status, gender, ethnicity, and so forth. They also examine the effects of differences in construct accessibility on the goals people define as salient for communication. In addition, echoing a common theme in this volume, they point to the importance of considering the influence of interpersonal communication on social cognition.

Housel offers an analysis of the central components of communicative interactions that any explanatory theory must include. He reviews the differences between conceptualizations of context, topic/content management, inference-making processes, memory, and conversational structures in the organization of communicative interactions. Finally, he explores "schema theory" implications for the analysis of conversations.

Following these two general analyses of the organization of adult communication, Kraut and Lewis offer research examining the effects of explicit and "back-channel" feedback on conversational situations and outcomes. They focus on two specific functional contexts for communication: situations in which deception is the primary goal, and those in which the goal is the summarization and transmission of information. In both contexts they find that differences in the nature of feedback clearly influence the structure and content of the speaker's messages, as well as the conversational outcomes. Their research provides a springboard for future analysis of the multiple functions of feedback in conversation. It also provides guidelines for exploring further the relationships among conversational goals, motives, prior knowledge, and feedback.

The final section of the volume contains two essays addressing the central role of social cognitive and communicative processes in the negotiation of relationships. Each points to important inadequacies in current thinking about relationship development involving matching, filtering, and/or social penetration models. O'Keefe documents the central role of relational and task goals in determining the quality of impression development over time. Her analysis clearly suggests that one's impressions of relational partners need not, and in fact often do not, proceed to increasing depths of knowledge about the character of another. Instead, person information is organized around particular goals perceived for the relationship, and the role of particular aspects of social cognitive structure in the impression formation process will vary depending on those goals.

Duck, Miell, and Miell critique positions that treat variables such as perceived similarity, self-disclosure, liking, empathy, and so on as determinants and/or natural outcomes of relational development. Rather, they argue that these factors are best seen as products of the communicative strategies partners employ. The authors call for greater attention to the influence of contextual and individual differences on the quality of social cognition and strategic processes in relationships. Their analysis of the organization of strategic processes in relationship growth and decline suggests the importance of grounding relational studies within the general study of social cognitive and communicative processes.

Delia concludes with a consolidation of the central themes running explicitly and implicitly through the volume. These themes reflect current trends in work on social cognition and communication: the move to more differentiated and specific measures of social cognitive functions; the development of more refined analyses of communication behavior that specifically acknowledge the centrality of goals and contexts in the organization of communication; and the treatment of communicative goals and strategies as independent phenomena influencing social cognition. Delia places these trends in historical perspective and points to their implications for future theory and research in the area.

These themes are common challenges confronting researchers in their attempts to explain how children and adults comfort, persuade, and establish relationships with others. A coherent account of such interpersonal processes will mean abandoning the limitations imposed by traditional disciplinary and topical boundaries. From the outset, we saw this volume as an effort to cross those boundaries and make our common concerns more apparent.

PART I

DEVELOPMENTS IN CHILDHOOD: PROSOCIAL PROCESSES

1

The Development of Children's Prosocial Cognition:
Research, Theory, and New Perspectives

Nancy Eisenberg and Rainer Silbereisen

In recent years there has been a dramatic increase in the quantity of empirical research on the development of prosocial behavior in children. A variety of topics have been examined, including the demographic, personality, and cognitive correlates of prosocial development and the role of culture, socialization practices, and other environmental influences on prosocial development (Eisenberg, 1982a; Rushton, 1980; Mussen & Eisenberg-Berg, 1977; Staub, 1979).

In many cases researchers who have studied positive behaviors have not derived their hypotheses from major psychological theories, but have been guided by the "quantification paradigm." According to this approach, the primary purpose of research is to obtain descriptive data not directly related to any particular theoretical orientation (Eckensberger, 1979, 1981). When in-

Authors' Note: *This research was supported by grants to the first author from the Provost Research Incentive Fund at Arizona State University and from the German Academic Exchange Service.*

vestigators interested in the development of prosocial behavior have based their research on the assumptions of a particular theory, the theoretical orientation has usually been either social learning theory or cognitive developmental theory (Mussen & Eisenberg-Berg, 1977; Rushton, 1980).

In the United States, social learning approaches clearly have had the strongest impact. This theoretical orientation sees behavior as shaped by the environment via the mechanisms of reinforcement, punishment, and observational learning. The large number of studies in which the role of these mechanisms in the development of prosocial behavior is examined attests to the theory's impact. Although not all researchers who have studied these mechanisms acknowledge the theoretical roots of their work, the influence of social learning concepts is evident.

Social learning theory has shaped the research concerning social behavior in yet another way. Altruistic behavior (the type of prosocial behavior that is of most interest to researchers) generally has been defined as intentional, voluntary behavior motivated by the desire to benefit another rather than by the desire for external rewards (Bar-Tal, 1976; Eisenberg, 1982c; Krebs, 1970; Mussen & Eisenberg-Berg, 1977; Staub, 1979). Despite the undeniable fact that definitions of altruism generally refer to internal cognitions (intentions, motives, and goals), until recently most researchers in the field have focused primarily on the development of visible behavioral manifestations of positive behaviors. Researchers' selective inattention to the cognitive components of prosocial development has been due in part to the traditional social learning view that overt behaviors, not internal processes, are the proper variables for study.

The instigation to examine the internal, cognitive aspects of prosocial development has derived primarily from the cognitive developmental perspective. The research and theoretical writings of both Piaget (1932/1965) and Kohlberg (1969, 1971) have stimulated interest in the role of moral judgment in prosocial as well as other types of moral behaviors (Dreman, 1976; Eisenberg-Berg, 1979b; Eisenberg-Berg & Hand, 1979; Rushton, 1975). Due to this impetus, researchers have also begun to discuss how changes in moral reasoning processes may be related to qualitatively different types of prosocial behavior and to age-related changes in an individual's predominant mode of prosocial responding (Bar-Tal, Raviv, & Leiser, 1980). Furthermore, as a response to Piaget's and Kohlberg's discussions of the roles of egocentrism and role taking in moral development, considerable attention has been given to the concept of role taking as a prerequisite or facilitator of prosocial behavior (Barnett, this volume; Hoffman, 1976, 1982; Iannotti, 1978; Kurdek, 1978; Mussen & Eisenberg-Berg, 1977; Staub, 1979; Underwood & Moore, 1982).

Despite the quantity of research that has been conducted from both a cognitive developmental and social learning orientation, there have been only a few attempts to conceptually integrate the development of the cognitive and behavioral components of prosocial behavior (see Bar-Tal, Sharabany, & Raviv, 1982; Hoffman, 1976, 1982; and Staub, 1979, for examples of these). Researchers occasionally have examined the degree of association between prosocial behavior and either moral reasoning or role taking (Blasi, 1980; Eisenberg, 1982b; Krebs & Russell, 1981; Kurdek, 1978; Underwood & Moore, 1982), but specification of the degree of association between prosocial reasoning and behavior does not clarify the processes underlying this relationship.

Due to the limited explanatory power of the traditional theoretical approaches in developmental and other areas of psychology, a new theoretical orientation has been gaining acceptance in Europe, especially in the German-speaking countries. This approach has been most frequently utilized and discussed by general psychologists (for example, von Cranach, in press; von Cranach & Harre, 1982; von Cranach, Kalbermatten, Indermuhle, & Gugler, 1982) and industrial psychologists (for example, Hacker, 1978; Volpert, 1974), but recently it has been applied to developmental issues (Eckensberger, 1979; Eckensberger & Reinshagen, 1980; Eckensberger & Silbereisen, 1980a, b), applied developmental and counseling research (Brandstader, 1981; Baumgardt, Kueting, & Silbereisen, 1981), and to the study of cross-cultural development (Eckensberger, 1979).

The action theoretical approach is quite complex. In reality, it comprises a group of related theories that are still evolving. Although action theory is new in that it has only recently been labeled as such and outlined in some detail (Eckensberger & Silbereisen, 1980a, b; von Cranach, in press), it has a diverse historical and philosophical basis. Action theorists have borrowed many ideas from Heider's (1958) *Naive Psychology*, especially thoughts about how people in everyday life develop elaborate ways of thinking about actions and their causes. From sociology (for example, symbolic interactionism), psychologists have borrowed the notion that manifest behavior is governed by cognitions which themselves derive from society (Goffman, 1963, 1969). Further intellectual input has come from industrial psychology (Hacker, 1978; Volpert, 1974) and from work on sytems theory and problem-solving (Miller, Galanter, & Pribram, 1960).

Among the major propositions of an action theoretical orientation is the assertion that human beings are self-reflective and that their actions are guided by mental representations. Thus, concepts such as goals, intentions, self-monitoring during action, and self-regulation by feedback cycle are fundamental aspects of the theory. Furthermore, society is an essential element in action theoretic formulations. According to an action orientation, the individual always acts in the environment, so that culture influences behavior via the specific circumstances surrounding an action. Moreover, society's rules govern the individuals' cognition and actions, and individuals in turn shape their culture by means of the consequences of their actions.

THE VON CRANACH MODEL

One of the better-known action theorists, Mario von Cranach, has represented the basic action theorem underlying an action-theoretical approach as follows: "In goal-directed action (in the context of acts), manifest behavior is governed by (partly) *conscious cognitions*, which in turn are (partly) of a social origin, so that society (partly) creates and controls the individual's action by controlling his cognitions, while the individual, by means of his acts, brings societal patterns into existence" (von Cranach, in press). In his formulation, the term *goal-directed action* (or simply *action*) refers to an actor's "consciously *goal-directed, planned* and *intended behavior,* which is *socially directed and controlled*" (emphases in original).

More specifically, there are three components to action. These three components refer to mental representations of alternative goals and outcomes, the attractiveness of those representations, and the mobilization of effort (Ecksberger & Silbereisen, 1980a). Moreover, action is organized along two dimensions, sequence and hierarchy. Sequence refers to the temporal patterning of behavior units at a given level of organization; hierarchy refers to super- and subordinate patterns of organization. The three principal levels of heirarchical organization are the goal level, where behavior is structured into acts by the cognitive determination of socially acceptable goals; the strategic level, where the course of action is cognitively steered by structuring sequences of action steps; and the operational level, where action steps are organized in their structural details by means of self-regulation (von Cranach, in press). Thus, in an action theoretic approach, cognition is intimately involved in the choice of actions, and in the organization and execution of action.

Von Cranach has delineated three classes of concepts, all of which contribute to an action structure. These are the manifest behavior, social meaning, and conscious cognition of the actor. According to von Cranach's model, manifest behavior and conscious cognition influence one another reciprocally, while the social meanings of actions to members of society affect one's conscious cognition. Von Cranach suggests that these three inputs should all be measured since they all contribute to action: manifest behavior (with systematic observation; see von Cranach, in press, for an example); conscious cognition (with interview data and content analyses of these data); and social meaning (by performing content analyses on individuals' naive interpretations of actions and actors). Thus von Cranach emphasizes that individuals' cognition about their own and others' behaviors is an essential element for the understanding of action.

Action theoretic formulations may provide a fertile basis for the integration of both conceptualizations and empirical research concerning prosocial development. The theory provides a framework for examining the multiple roles of cognition in choosing and executing prosocial acts, in addition to perspectives concerning the complex interplay of societal factors and individuals' behavior. The role of cognition in an action theoretic perspective, and the methodological implications of this perspective, will be discussed at the end of this chapter. First, research regarding prosocial cognition will be summarized.

EMPIRICAL RESEARCH ON
THE DEVELOPMENT OF
PROSOCIAL COGNITIONS

The three classes of inputs that von Cranach included in his model—manifest behavior, social meaning, and conscious cognition—provide a useful framework for organizing the research on the reasoning behind prosocial behavior. The concept of manifest behavior corresponds to overt prosocial behavior and thus will not be discussed further in this section of the chapter. The second catetory, social meaning, refers to how members of society interpret actions; that is, to people's naive interpretations (attributions) of others' social behavior, including prosocial actions. Much of the research on prosocial cognition fits into this category—specifically, research on people's attributions regarding the kindness of others' behaviors. Finally, the third type of variable that von Cranach considers important for understanding action— people's conscious cognition—corresponds to research in which people who have behaved in a prosocial manner are interviewed regarding their motives and reasoning in that situation.

Additional research explores individuals' conscious cognition, but more indirectly. This is the research on children's reasoning regarding hypothetical moral dilemmas about helping actions. In this research, the children themselves are not actors. However, their cognitions about the hypothetical conflicts presumably reflect how children reason about their behavior when they are confronted with a difficult moral decision.

The research regarding the development of children's attributions about prosocial actions and their conscious cognition concerning prosocial actions will now be reviewed.

Social Meaning: Children's Attributions
Regarding the Kindness of Actions and Actors

The type of methodology most frequently used by researchers to examine children's naive interpretations of others' positive behavior is a modification of one of the procedures used by Piaget (1932/1965) to study children's concepts of morality. Piaget frequently elicited children's attributions (and the reasoning behind their attributions) by presenting two or more scenarios involving moral (or immoral) actors, asking the children which story protagonist was naughtier or which story solution was the fairest, and then requiring the children to justify their decisions. The children were asked to make attributions regarding actors or actions, and this information was used as the basis for conclusions about how children view moral issues and actors. Attributional data of this sort can be used to measure the social meaning of various prosocial actions.

Clara and Alfred Baldwin were two of the first to adapt Piaget's methods to study people's attributions regarding the kindness of people and their actions (Baldwin & Baldwin, 1970; Baldwin, Castillo-Vales, & Seegmiller, 1971). They defined kindness as "a motivation that is sometimes inferred from the fact that one person benefits another, provided the circumstances are appropriate" (Baldwin & Baldwin, 1970, p. 30). Some of the circumstances that they hypothesized would influence an observer's judgments of the kindness of an actor's beneficial actions were the intentionality of the act; whether or not the benefactor had a choice; whether or not the benefactor was acting in obedience to a request or command from authority; whether or not the benefactor was acting in his or her own self-interest (for example, helping in an attempt to bribe another or to promote trade); and whether or not the benefactor was acting in accordance with a social obligation. The Baldwins hypothesized that an action that benefited another would be viewed as less kind by an observer if the act was not executed by choice or was not intentional, or if it benefited the actor or fulfilled a social obligation.

In general, the Baldwins' research supported their predictions. Among adults, an act was judged as kinder if it was intentional, voluntary, involved self-sacrifice, did not benefit the actor, and did not fulfill a social obligation. Moreover, there were developmental changes in children's attributions concerning the kindness of an act. Second graders' attributions regarding intentionality were like those of adults, as were their evaluations of the kindness of an actor who assisted to promote a trade. By fourth grade, children agreed with adults in their inferences regarding the role of choice, obedience to authority, self-sacrifice, or obligation to a guest. By eighth grade, the children were similar to adults in most attributions regarding kindness, including attributions regarding the role of returning a favor. Only the evaluation of one type of social obligation, the equalization of benefits, showed further development from eighth grade to adulthood.

In a more recent study, Leahy (1979) examined the attributional schemes used by children and adults to make inferences regarding an actor who helps another. Specifically, Leahy tested for developmental changes in children's use of discounting, augmentation, and additive principles (Karniol & Ross, 1976, 1979; Kelley, 1967, 1972). According to the discounting principle, a given cause (for example, the assumption that an act was motivated by kindness) is minimized when another plausible facilitory cause (such as a formal obligation or duress) is present. In contrast, according to the augmentation principle, the presence of an inhibiting cause (for example, a previous refusal to help by the potential recipient of aid, or threats of physical harm to the potential benefactor if he or she helps) leads to the inference that the facilitory cause (such as the kindness of the actor) is greater. Finally, according to the additive principle, an actor's intentions are believed to be positively related to the outcome of his or her behavior. Thus, if a person is rewarded for helping, he or she is perceived as wanting to help more (and, consequently, as being more kind) than a person who is not rewarded.

Leahy found that young children (but not sixth graders or adults) viewed an act as kinder if it was rewarded than if it was not rewarded (that is, they used the additive principle). In contrast, in accordance with the discounting principle, fifth graders and adults also judged actors who assisted in spite of the threat of harm as more kind than actors who assisted when doing so resulted in the avoidance of physical harm (using the augmentation principle rather than the discounting principle). Unlike first or fifth graders, adults believed an individual to be more kind if he or she helped when there was a reciprocity-related reason not to assist—for example, when the potential recipient had not shared with the potential benefactor on a previous occasion. Finally, subjects of all ages viewed story protagonists as more kind if they were consistent in their behavior and helped more than one individual.

DiVitto and McArthur (1978), like Leahy, examined developmental changes in children's and college students' attributional processing. The subjects were asked to evaluate both the agent of the act and the target in hypothetical stories about prosocial, negative, and neutral behavior. Even DiVitto and McArthur's first graders could infer that an actor who consistently shared with another was kinder than one whose history of sharing was less consistent. Furthermore, they were able to infer that an actor who shared with many people was kinder than one whose sharing was limited to one recipient. However, the first graders did not infer that a target with whom most people shared was nicer than one with whom almost nobody shared. The ability to do so increased with age, and performance was better than chance by third grade. Moreover, use of the discounting principle increased with age, while reliance on the additive principle decreased with age (significantly between grades three and six, and between grade six and adulthood).

Several other researchers who have examined children's attributions regarding prosocial behaviors have obtained data similar to those already discussed. In general, children's use of the additive principle has been found to decrease significantly by second or third grade. Conversely, use of the discounting or augmentation principles with regard to both prosocial behavior (Benson, Hartmann, & Gelfand, 1981; Cohen, Gelfand, & Hartmann, 1981; Petersen, 1980; Shure, 1968; Suls, Witenberg, & Cutkin, 1981) and other behaviors (Costanzo, Grumet, & Brehm, 1974; Karniol & Ross, 1976, 1979) apparently increases with age. Use of the discounting principle seems to increase significantly during the early school years until approximately third grade, and gradually thereafter. While use of the discounting principle usually is not reliable before age seven to nine, younger children (for example, kindergartners) will discount an actor's internal motivation when they are helped to focus on and understand the inducement value of interpersonal causes (Benson et al., 1981; Karniol & Ross, 1979; Kassin, 1981).

It is interesting to note that children's attributions regarding others' behavior may differ in systematic ways from their attributions regarding their own behavior. For example, it appears that children are more likely to discount others' prosocial motives than their own (Smith, Gelfand, Hartmann, & Partlow; 1979; Cohen et al., 1981). Gelfand and Hartmann (1982) suggest that this pattern of findings is due to motivational factors; that is, that self-image maintenance and self-esteem are better served by attributing one's own prosocial behavior to altruistic motives rather than to hedonistic materialism.

Conscious Cognition: Children's Moral Judgments About Prosocial Behavior

In research on moral judgment, individuals are not asked to make attributions regarding the kindness of another actor's behavior. Rather, subjects are asked to do one of two tasks. In some studies an individual is questioned regarding the reasons for his or her own prosocial behavior. In this type of research, individuals' attributions regarding their own motivational and reasoning processes are elicited.

In the second type of moral judgment research, people are presented with hypothetical moral dilemmas. They are then asked to resolve dilemmas and to explain their reasoning processes. With some exceptions (for example, Eisenberg-Berg & Neal, 1981), subjects are usually told to make a judgment regarding a hypothetical third person's actions. Nevertheless, it is generally assumed that the responses elicited from a subject reflect, at least to some degree, his or her own reasoning in those situations in which the individual actively attempts to resolve a moral dilemma involving an identical (or similar) conflict of values.

The two different methodologies used to assess people's reasoning about prosocial acts provide us with different kinds of information, and each has its strengths and weaknesses. On the one hand, when people are questioned regarding their own real-life prosocial behavior, their responses should have more ecological validity than if they were reasoning about a hypothetical situation. They should have better access to their own thought processes than if they were reasoning about a hypothetical situation, and they should be reflecting on their own motives and reasoning rather than on how they believe that others might reason. In contrast, when reasoning about a hypothetical dilemma, people must speculate as to how they themselves or some other person would reason in a situation that may not seem real.

On the other hand, there may be substantial problems with the interpretation of individuals' post facto reasoning about their own behavior. It is quite possible that people frequently alter their thinking (consciously or unconsciously) to justify previous actions, or that they construct new attributions concerning their behavior and value systems based on reflection about their own previous behavior (Bem, 1972). In either case, distortion of the cognitive processes that originally instigated or were involved in the execution of an action may occur. When individuals reason regarding hypothetical moral dilemmas, these types of distortions should not occur, although contemplation of one's own past behavior and the desire to distort one's own reasoning processes certainly may influence an individual's moral judgments.

Given the weaknesses of the two methodologies commonly used to assess an individual's conscious cognitions regarding prosocial actions and moral decision making, the most circumspect approach to interpreting the empirical data is to look for consistencies in the results of research conducted with these (and other) methodologies and to weight consistent results more heavily than less consistent findings. Of course, systematic differences in the patterns of results from the two different approaches would also be of interest. Such differences could provide clues regarding the role of other affective and cognitive processes that intervene in and/or alter the relationship between value-related thinking and the execution of a behavior.

Developmental changes in reasoning regarding children's own prosocial behaviors. In most research concerning children's reasoning about their own prosocial behavior, children either have been given the opportunity to assist another or have been induced to help or share and were then interviewed regarding their subsequent behavior. One of the earliest examples of this type of study is that of Ugurel-Semin (1952). After each of 291 Turkish children aged four to sixteen had been asked to divide an uneven number of nuts between himself or herself and a peer, the children were questioned regarding their sharing behavior. Ugurel-Semin found that the youngest children reasoned in an egocentric manner; that is, they exhibited a purely selfish orientation, or they confused the self with others or the material with the moral. Children of about nine years reasoned primarily in terms of obedience to stereotypic, socially accepted rules and norms (concerning when and with whom one shares), and in terms of shame for violation of these rules. In contrast to younger children, children of about ten years or older emphasized the importance of maintaining positive interpersonal relationships, empathic reasons for sharing, and internalized values regarding just behavior.

Other researchers have found that young children's reasoning regarding their own prosocial behavior may be less egocentric than Ugurel-Semin's data would indicate. Damon (1971) found that preschool children's reasoning regarding why they shared with a peer was frequently empathic (indicating concern about the consequences of their behavior for others), as well as pragmatic and self-oriented (concerned with the reciprocal consequences of sharing for the friendship or for oneself). Similarly, in research with kindergarten-aged children, Dreman and Greenbaum (1973) found that children justified previous sharing with a classmate on the basis of (in order of importance from most to least) norm-directed reasons (the child felt obligated to share with a dependent other because of prevailing social norms); empathic reasons (stemming from the desire to make the recipient happy; no interpersonal profit or loss was supposedly involved); and interpersonally oriented reasons (the child wanted to perpetuate existing friendships or repay friends for past services). Although the interpersonal reasons sometimes seemed to reflect egocentric, self-oriented goals, the children frequently verbalized more altruistic motives.

Bar-Tal and his colleagues in Israel have examined children's conscious congnition about their own prosocial actions in a variety of different situations. In their experimental paradigm, children are usually presented with a series of opportunities to assist another, proceeding from the least explicit (with regard to pressure and/or the offer of concrete rewards) to the most explicit. When the children do assist, they are questioned regarding their motives for doing so. For example, in one recent study (Bar-Tal et al., 1980) children played a game with a peer, won candy, and then were provided with a series of opportunities to share the candy with the peer (the sequence of opportunities ended when the child did indeed share).

In Bar-Tal's studies, the children's reasoning about their motives was coded into various categories. These are similar to those used by Bar-Tal et al. (1980) and by Raviv, Bar-Tal, and Lewis-Levin (1980):

(1) Concrete reward—prosocial behavior occurred because of a promised reward.
(2) Compliance—prosocial behavior occurred because the experimenter told the child what to do.
(3) Internal initiative with concrete reward—prosocial behavior occurred because the child believed that he or she would receive a concrete reward for performing the act.
(4) Normative—prosocial behavior occurred because of a normative belief prescribing sharing with other children. Conformity with this norm brings social approval (for example, "It's nice to share").
(5) Generalized reciprocity—prosocial behavior occurred because of a belief in a generalized social rule that people who act prosocially will receive aid when they are in need.
(6) Personal willingness with external reward, but also with expressions of self satisfaction (for example, "I like to share to give others satisfaction").
(7) Personal willingness to act prosocially without any reward (for example, "Candy should be shared to make the other children happy").

In general, Bar-Tal and his colleagues have found that normative reasoning is used most frequently by both kindergartners and school-aged children to justify their actions. Children, especially kindergartners, also frequently justify their actions with references to concrete rewards. Sharing due to more altruistic, internal, or empathic reasons occurs relatively infrequently, although it increases somewhat with age.

Although Bar-Tal and his colleagues seem to view the types of reasoning listed above as components of a developmental sequence, the research concerning this issue is limited. In a study involving fourth, sixth, and eighth grade children, mode of reasoning was only marginally related to age (Raviv et al., 1980). However, in two other studies involving kindergarten and elementary school children up to the fourth grade, the predicted age-related patterns of use were found (Bar-Tal et al., 1980; Guttman, Bar-Tal, & Leiser, 1979).

Unfortunately, even when Bar-Tal and his colleagues have obtained the predicted relationship between type of reasoning and age, it is unclear whether the stages actually developed in the sequence delineated, or if the relationships with age were due to age changes in the use of only a few of the reasoning categories. It is also difficult to examine developmental change with Bar-Tal and his colleagues' data because the children in any given study frequently were reasoning about very different situations. Recall that the children were questioned regarding their motives only *after* they shared. Since children shared under different circumstances (at different points in the sequence of sharing opportunities), and since age was related to the type of situation in which the children shared, any developmental changes in the children's reasons could have been due to the differences in the circumstances under which they shared (indeed, reasoning level was related to the type of situation in which the children shared first). Finally, it should be noted that some of the difficulties in interpreting Bar-Tal and his colleagues' data are due to the fact that they were not focusing primarily on questions relating to the children's reasoning regarding their own motives.

In the research discussed above, the prosocial behavior that the children discussed occurred in response to a contrived circumstance. Using a slightly different methodology, Eisenberg-Berg and Neal (1979) examined children's reasoning about their own naturally occurring prosocial behavior rather than artificially elicited behavior. In this research, preschoolers were casually interviewed by a familiar adult whenever the adult saw them perform a peer-directed prosocial act. Under these circumstances, the children most frequently explained their prior prosocial behavior by referring to the needs of others or to pragmatic concerns. References to rewards, punishment, approval, or norms were relatively infrequent.

As demonstrated by this review of the literature, there are both consistencies and inconsistencies in the research concerning children's conscious cognitions about their own prosocial behavior. The inconsistencies are undoubtedly due in part to the fact that the children in the studies reviewed above were from three different cultures, and that the situations in which the children assisted another varied dramatically across studies. In some of the studies there were adults present when the children assisted; in others, there were no adults observing the children. Furthermore, in some of the research, especially that of Bar-Tal and his colleagues, the children did not share or help until they were directed to do so or were told that they would be rewarded for assisting. In most of the other research, the children were not explicitly told to share and were not offered concrete rewards for assisting another; thus, the children's choices were less constrained. Finally, in some studies other situational variables were being manipulated so that not all subjects within a particular study were reasoning about sharing under the same conditions (for example, Bar-Tal et al., 1980; Dreman & Greenbaum, 1973).

Despite the inconsistencies in the research paradigms and data, it is possible to draw some tentative conclusions regarding children's conscious cognitions about their own prosocial actions. In general, preschoolers' and elementary school children's reasoning seems to be predominantly pragmatic, empathic, and self-oriented (including being oriented to reciprocity concerns and to the desirability of assisting someone important to the child) (Bar-Tal et al., 1980; Damon, 1971; Dreman, 1976; Dreman & Greenbaum, 1973; Eisenberg-Berg & Neal, 1979; Guttman et al., 1979; Smith et al., 1979; Ugurel-Semin, 1952). Normative reasoning seems to be expressed by young children primarily when an adult is either present during the prosocial act or instigates it. Perhaps the presence of an adult increases the probability that a child will focus on normative standards that are part of adult society. In none of the studies did children frequently say that their prosocial behavior was motivated by fear of punishment or blind obedience to an authority's dictates (reasoning analogous to Kohlberg's [1969] Stage 1), even if an adult had suggested that the child should share or help.

From age four to elementary school age, the amount of reward-oriented reasoning used by children appears to decrease (Bar-Tal et al., 1980; Ugurel-Semin, 1952). Simultaneous with the decrease in reward-oriented reasoning, other-oriented, altruistic reasoning—indicating a willingness to share without the expectation of external rewards, and frequently reflecting self-reflective role taking and/or empathic responses or an emphasis on equality and justice—increases in frequency (Bar-Tal et al., 1980; Bar-Tal et al., 1981; Ugurel-Semin, 1952). In general, the pattern of data is consistent with the conclusion that as children develop, the types of cognition instigating, controlling, or associated with prosocial actions become more internal and less related to external gain.

Children's moral judgments about hypothetical moral dilemmas. Much of the research on children's moral judgments about hypothetical moral dilemmas has been modeled after the research of Kohlberg (1969, 1971). Kohlberg devised a methodology in which subjects are presented with a series of stories, each of which involves a moral conflict. Subjects are requested to resolve the moral conflict and to explain their reasoning. The reasoning used by the subject, rather than his or her choice of story resolution, is the unit of analysis.

Various researchers who have adapted Kohlberg's methodology have attempted to delineate the reasons used by people to resolve moral dilemmas in which one possible outcome is prosocial behavior. Thus the decision-making process in resolving conflicts of values and/or needs is examined, not the criteria and processes used to make inferences regarding one's own or others' behavior. Perhaps more than any other, Eisenberg has explicitly differentiated between prohibition-oriented moral reasoning and prosocial moral reasoning. In her view, Kohlberg's moral dilemmas deal with primarily prohibition-oriented issues; that is, issues relating to rules, laws, various authorities' dictates, punishment, and formal obligations. Although prosocial action is one possibility in several of Kohlberg's dilemmas, potential prosocial behavior in these situations would result in the violation of a formal prohibition. For example, in Kohlberg's Heinz dilemma, Heinz can save his wife's life, but only if he breaks the law by stealing. With Kohlberg's hypothetical moral dilemmas, one cannot examine reasoning about prosocial behavior in situations where one person's needs conflict with those of another but in which no authorities or formal prohibitions exist to direct the individual's course of action.

In one series of studies, Eisenberg and her colleagues and students examined what she labels prosocial moral reasoning; that is, reasoning about conflicts in which the individual must choose between satisfying his or her own desires or needs and those of others in a context where the role of laws, punishment, authorities, formal obligations, and other external prohibitions is irrelevant or deemphasized (Eisenberg, Lennon, & Roth, 1982; Eisenberg-Berg, 1979a, b; Eisenberg-Berg & Hand, 1979; Eisenbeg-Berg &

Neal, 1979; Eisenberg-Berg & Roth, 1980; Mussen & Eisenberg-Berg, 1977).

In a cross-sectional study involving second, fourth, sixth, ninth, eleventh, and twelfth graders, Eisenberg (Eisenberg-Berg, 1979b; Mussen & Eisenberg-Berg, 1977) examined developmental changes in children's prosocial moral judgments. The children's reasoning responses were coded into a variety of different moral judgment categories, many of which were similar to aspects of Kohlberg's (1969, 1971) stages. The types of reasoning that were related and that showed the same developmental trends were grouped together into age-related "stages" or levels of reasoning. These levels, modified slightly as a result of longitudinal research with young children (Eisenberg-Berg & Roth, 1980), are presented in Table 1.1. According to these age-related levels, hedonistic, self-oriented reasoning is the least developmentally mature type of reasoning, followed by needs-oriented reasoning (primitive empathic reasoning), stereotyped and interpersonal approval-oriented reasoning, and then by self-reflective, overtly empathic reasoning. The highest level of reasoning includes judgments based on internalized values, norms, or duties, and on guilt or positive affects relating to the maintenance of self-respect by living up to one's own internalized values. The less mature types of reasoning tend to decrease in frequency with age. Nevertheless, childhood modes of reasoning were verbalized by even the oldest of adolescents, especially when they were justifying the decision not to assist a needy other.

In subsequent longitudinal research in which children were interviewed at ages of approximately four to five, five to six, and seven to eight, Eisenberg further examined the early development of reasoning about prosocial dilemmas. She found that need-oriented (other-directed) reasoning increased with age from the preschool to school years (Eisenberg-Berg & Roth, 1980), while hedonistic reasoning simultaneously decreased in frequency. This same pattern of change continued during the early school years (Eisenberg et al., 1983).

In general, the results of the research on children's reasoning regarding hypothetical moral dilemmas are consistent with the findings concerning children's cognition about their own prosocial behavior, at least for preschool and elementary school (there is little research on adolescents' cognition about their own behavior). This consistency seems to be greater if the circumstances in the real-life situation and the hypothetical situation are similar on

TABLE 1.1   Levels of Prosocial Moral Reasoning

1. Hedonistic, self-concerned orientation: The individual is concerned with selfish, pragmatic consequences rather than moral considerations. "Right" behavior is that which is instrumental in satisfying the actor's own needs or wants. Reasons for assisting or not assisting another include consideration of direct gain to the self, future reciprocity, and concern for others because the individual needs and/or likes the other.

2. "Needs of others" orientation: The individual expresses concern for the physical, material, and psychological needs of others, even though these conflict with his or her own needs. This concern is expressed in the simplest terms, without clear evidence of role taking, verbal expressions of sympathy, or reference to internalized affect such as guilt ("He's hungry" or "She needs it").

3. Approval and interpersonal orientation and/or stereotyped orientation: Stereotyped images of good and bad persons and behavior and/or considerations of others' approval and acceptance are used in justifying prosocial or nonhelping behaviors. For example, one helps another because "it's nice to help," or because "they'd like him more if he helped."

4a. Empathic orientation: The individual's judgments include evidence of sympathetic responding, role taking, concern with the other's humanness, and/or guilt or positive affect related to the consequences of one's actions. Examples include "He knows how he feels," "She cares about others," and "She'd feel bad if she didn't help because they'd be in pain."

4b. Transitional stage: Justifications for helping or not helping involve internalized values, norms, duties, or responsibilities, or refer to the necessity of protecting the rights and dignity of other persons; these ideas, however, are not clearly and strongly stated. References to internalized affect, self-respect, and living up to one's own values are considered indicative of this stage if they are weakly stated. Examples include "It's just something I've learned and feel."

5. Strongly internalized stage: Justifications for helping or not helping are based on internalized values, norms, or responsibilities, the desire to maintain individual and societal contractual obligations, and the belief in the dignity, rights, and equality of all individuals. Positive or negative affects related to the maintenance of self-respect and living up to one's own values and accepted norms also characterize this stage. Examples include "She'd feel a responsibility to help other people in need," and "I would feel bad if I didn't help because I'd know that I didn't live up to my values."

SOURCE: Adapted from Eisenberg (1982b).

some critical dimensions (for example, whether or not there are adults directly involved, whether or not the relationship of the potential recipient to the actor is similar, and whether or not concrete rewards are offered to the actor for assisting).

As was true with regard to children's reasoning about their own behavior, children tend to resolve hypothetical moral dilemmas primarily with self-oriented or simple empathic reasoning. With age, they generally become more attuned to interpersonal considerations. However, if the story characters in the hypothetical dilemma had been friends or relatives, it is likely that the children would have exhibited more interpersonally oriented reasoning at an earlier age, as was found by Dreman and Greenbaum (1973). Self-reflective empathic justifications also become more frequent with age, as does reasoning concerning internalized values and norms, and affects relating to the consequences of one's behavior for others or to living up to one's own values. In brief, the results of research on both children's moral reasoning about their own behavior and their reasoning about third-person hypothetical dilemmas suggest that an individual's thinking regarding prosocial action becomes more complex, internal, and self-reflective with age.

Interpretation of the Data Resulting from the Various Methodologies

Thus far, three different types of cognition about prosocial behavior have been discussed: attributions regarding the kindness of another actor or another's actions, reasoning about one's own prior prosocial behavior, and reasoning concerning hypothetical moral conflicts. It is logical to assume that each of these categories of cognition provides information regarding cognitive processes that might influence individuals' prosocial actions, but that this information may be of a somewhat different nature for each category. It is likely that the results of research on attributions about others' actions will contribute to our understanding of prosocial action in at least two ways. First, as mentioned previously, analyses of this type of data can provide information regarding the social meaning of the different types of positive behavior for various age groups in a society. In other words, if one considers the attributional data from a normative perspective, one can determine the behavior that people of a specific age view as kind.

Important insights into a child's social world can be derived from this type of data, especially concerning the social meanings held by the child's peers. Because it is likely that a child's values and overt behaviors are frequently influenced by the values and perceptions of his or her peers, it is important to determine how the age-mates in the individual's social milieu view prosocial behavior. Of course, the significance of a helping behavior to adults may also affect the manner in which a child reasons and behaves. Thus, the social meaning of a given prosocial act may derive from some combination of both peers' and adults' interpretations of the act.

It is reasonable to assume that not all children of the same age make inferences regarding prosocial behavior in an identical manner. Due to individual differences in cognitive abilities, some children will be more developmentally advanced than others with regard to attributional abilities. Furthermore, individual differences in values, needs, and other personal variables may influence the individual's inferential processes, regardless of his or her cognitive capacities. Consequently, the data on attributions provide more than information on normative interpretations of behavior at various ages; they are also a source of information regarding individual differences in attributional tendencies. Information of the latter type is of value because children's own evaluations of the kindness of an action (as well as the evaluations of other members of society) should influence why and when a child enacts a particular prosocial behavior.

In contrast to children's inferences regarding others' prosocial behavior, children's moral judgments about themselves and others would seem to reflect those values and motives of most importance to the individual. Furthermore, if the individual reasons about a variety of situations when he or she has to resolve a moral dilemma, the range of conscious considerations that might be used in potential helping situations should become clearer. These considerations may influence the individual's goals and, perhaps, the means by which he or she attempts to accomplish those goals.

More specifically, when children reason regarding their own prior prosocial behavior, it is logical to assume that their responses reflect some of the conscious motives and types of cognition that were operating in a specific situation, plus any conscious or unconscious distortions and other intervening cognitive or affective processes that might have occurred. Thus, embedded in the individual's post facto reasoning, there should be information regarding the end product of any cognitive conflict that proceeded or accompanied the prosocial act. Usually the child would not be expected to discuss value considerations that did not prevail in the specific situation when justifying a prior course of action. In contrast, children's moral reasoning about hypothetical situations would seem to provide insights regarding the values that the child typically considers before engaging in prosocial behavior (or before deciding not to help in a specific situation). Since no pro-social behavior is actually performed by the child when he or she reasons about hypothetical moral dilemmas, the hypothetical moral judgment data reflect the range of motives and type of cognition that might be brought to bear upon a situation, not just the reasoning that finally prevails. Furthermore, by considering the relative influence of various types of conscious cognitions in the decision-making process, it should be possible to determine not only which conscious values or goals are important to the child in resolving moral conflicts, but also the relative importance of these values and goals in their value hierarchy.

IMPLICATIONS OF AN ACTION THEORETICAL PERSPECTIVE FOR THE STUDY OF PROSOCIAL COGNITION

Perhaps one of the most positive consequences of contact with a new or different theoretical perspective is that it forces us to reevaluate old assumptions and the implicit models underlying our thinking. Moreover, it provides a different lens through which to filter the theory and data on the topic. Some of the issues highlighted by an action theoretic perspective will now be discussed.

Meaning of Action

As we discussed previously, prosocial behavior is usually de-fined as voluntary, intended behavior (Eisenberg, 1982c; Staub, 1979). Thus, researchers interested in prosocial behavior generally are concerned with a class of behavior that is likely to fall into the rubric of "action" (consciously goal-directed, planned, and in-tended behavior that is socially directed and controlled).

Action theorists emphasize that actions can occur only when an individual's behavior is not constrained; that is, when he or she has behavioral options and is free to choose among them. When in-dividuals' choices are highly restricted, his or her behavior is not intended and/or goal-directed, and such behavior does not fit the definition of action. This emphasis on the role of volition serves to underscore the involuntary nature of some types of behavior that are frequently labeled prosocial, and about which subjects are asked to reason. For example, in Bar-Tal and his colleagues' research (Bar-Tal et al., 1980; Raviv et al., 1980), children often were ordered to assist or share. Thus, if they did so, they were probably not acting in a prosocial manner (if prosocial behavior is defined as voluntary behavior intended to help another). Similarly, if helping behavior is elicited from all individuals in a situation, one has to question the degree to which such behavior is constrained by circumstances.

By constraining an individual's choices when he or she reasons about a moral dilemma, one is also losing valuable information concerning how people think about prosocial actions. For an action to be performed, the individual has to (1) attend to some subset of information in the environment relating to the prosocial act, and (2) continue to think about that subset of information rather than about any number of other matters or goals. In other words, if an individual is to actually behave prosocially in any given situation, he or she must both attend to this possibility and then cognitively pursue this course of action.

With story dilemmas or in constrained settings, individuals frequently are forced to attend to and contemplate prosocial behavior, even if they would not do so in real life. Thus, individuals' reasoning may not be representative of their thought processes in similar real situations. This is especially true for young children for several reasons: (1) in real-life situations, due to their distractability and limited experience, children may not attend to cues relating to prosocial behavior (Pearl, 1979); (2) children may interpret environmental cues differently from adults, and thus may not view a real-life situation in which an individual needs assistance as such; and (3) when children are not forced to focus explicitly on the needs of another (as is done somewhat with story dilemmas), other types of cognition, especially those stemming from self-oriented needs, may be more likely to monopolize and/or intrude upon a child's thinking. Thus it is important for researchers in the future to examine how people interpret the larger context in which the potential for prosocial action is embedded (see, for example, Barnett, Darcie, Holland, & Kobasigawa, 1982). This is necessary if we are to determine what constitutes a moral conflict for individuals in different situations.

According to an action theoretic perspective, individuals' potential behavior is organized in a hierarchical structure based in part on a hierarchy of action goals (the imagined state inspired as the outcome of an action) and superordinated goals. Behaviors that are relatively low in this hierarchy (because the goals that underlie the behaviors are relatively unimportant) are likely to be more routinized and may often seem trivial to the actor. Thus, the enactment of behavior low in the hierarchy should provoke less conscious thinking than would more important behavior. Indeed, if a particular type of prosocial behavior is relatively unimportant to an individual, he or she may frequently perform it nearly automatically (Karniol, 1982; Langer, Blank, & Chanowitz, 1978), due, perhaps, to the existence of cognitive scripts for unconflicted behavior. In such cases it should be difficult to elicit the individual's reasoning regarding such prosocial behavior, and any reasons that an individual reports may have been manufactured ex post facto. Only for behavior that is the result of active decision-making and/or planning might the individual be able to explain accurately his or her reasoning as it occurred. Consequently, researchers should not assume that people (especially children) actually reason about every positive act they perform, nor should they expect an association between individuals' general level of moral judgment and mechanically performed prosocial actions.

The behavior hierarchy referred to above differs among people. Particular kinds of helping behavior may be of more significance for some than for others. In order to understand developmental changes in prosocial responding, it would be helpful to know the degree of variation among individuals in the hierarchy of goals at different ages, and whether or not there are developmental trends in terms of the organization and structure of this hierarchy. Furthermore, information regarding individuals' hierarchies may help to explain inconsistencies between individuals' prosocial moral judgments and their prosocial behavior. This is because individuals may reason differently about different types of behavior (and thus behave in different ways) depending on their importance in the behavior hierarchy.

Implications for Research

The action theoretic perspective has several implications for the study of prosocial behavior and cognition. First, according to this perspective, both prosocial behavior and cognition should be considered and examined as part of an action sequence. More concretely, it is important for researchers to attend to the entire decision-making process, including which aspects of the situation

an individual attends to, how the individual processes and interprets information in a situation, and how the individual constructs
moral dilemmas. For example, if children are interviewed regarding their reasoning only when they behave prosocially, one loses
information regarding the decision-making process that leads
some children to fail to assist others.

A related issue concerns the role of the environment on an action sequence. Action theorists, like the social learning and
behavioral theorists in America, emphasize the fact that the larger
context provides the situation in which a behavior is embedded.
Contained in the situation itself are the possibilities or alternatives
that an individual can process, as well as the social influences that
might affect an action. Moreover, aspects of a particular situation
might either facilitate or inhibit the transformation of cognition
into performance. Thus, to fully understand prosocial cognition
and its role in action, one has to examine in detail the person-by-
situation interaction in the development and elicitation of prosocial
cognition.

How might this be done? For older children and adults, the
methods devised by von Cranach (in press) could be useful. He
asked individuals to perform meaningful acts under conditions as
natural as possible, videotaped their actions, and then, immediately afterward, played the videotape back to the actor. At this time
the actor was asked to report his or her remembered reasoning.
The actors' protocols were then analyzed. Moreover, in other
studies (see von Cranach & Kalbermatten, 1982; Kalbermatten &
von Cranach, in press) researchers have shown naive observers a
videotape of ongoing action and asked them to interpret what they
saw. Similar techniques would be useful for the study of reasoning regarding prosocial behavior.

With younger children, verbal techniques of this sort would be
less appropriate, although children could interpret short tapes of
their own and others' behavior. Moreover, one could learn about
young children's thinking regarding prosocial acts by question-
ing children about how they would process and react to real-life
or detailed hypothetical situations in which the potential for prosocial action was embedded, and by eliciting children's reasoning regarding their decisions when they decided not to help, as
well as when they assisted. Analyses of videotapes of children's
naturally occurring behaviors may provide clues as to how children
process information relating to potential prosocial actions.

In summary, although we now have some data relating to the
development of prosocial cognition, these data are limited in

scope. They do not contain information on all aspects of the role of cognition in the decision-making process. Methodologies in which the entire action sequence is examined, especially in natural circumstances, could greatly enhance our understanding of how and when prosocial moral cognition is constructed.

REFERENCES

Baldwin, A. L., Baldwin, C. P., Castillo-Vales, V., & Seegmiller, B. (1971). Cross-cultural similarities in the development of the concept of kindness. In W. W. Lambert & K. Weisbrod (Eds.), *Comparative perspectives in social psychology.* Boston: Little, Brown.

Baldwin, C. P., & Baldwin, A. L. (1970). Children's judgments of kindness. *Child Development 41,* 29-47.

Barnett, K., Darcie, G., Holland, C. J., & Kobisagawa, A. (1982). Chidren's cognitions about effective helping. *Developmental Psychology, 18,* 267-277.

Bar-Tal, D. (1976). *Prosocial behavior: Theory and research.* New York: Wiley.

Bar-Tal, D., Raviv, A., & Leiser, T. (1980). The development of altruistic behavior: Empirical evidence. *Developmental Psychology 16,* 516-525.

Bar-Tal, D., Raviv, A., & Shavit, N. (1981) Motives for helping behavior: Kibbutz and city children in kindergarten and school. *Developmental Psychology, 17,* 766-772.

Bar-Tal, D., Sharabany, R., & Raviv, A. (1982). Cognitive basis of the development of altruistic behavior. In V. Derlega & J. Grzelak (Eds.), *Cooperation and helping behavior: Theories and research.* New York: Academic Press.

Baumgardt, C., Kueting, H. J., & Silbereisen, R. K. (1981, September). *Social competence as an objective of road safety in education in primary schools.* Paper presented at the International Conference on Road Safety, Cardiff, Wales.

Bem, D. J. (1972). Self-perception theory. In L. Berkowitz (Ed.), *Advances in experimental social psychology* (Vol. 6). New York: Academic Press.

Benson, N. C., Hartmann, D. P., & Gelfand, D. M. (1981, April). *Intentions and children's moral judgments.* Paper presented at the Biennial Meeting of the Society for Research in Child Development, Boston.

Blasi, A. (1980). Bridging moral cognition and moral action: A critical review of the literature. *Psychological Bulletin, 88,* 1-45.

Brandstader, J. (1981, September). *Entwicklungs-beratung unter dem aspekt der Lebensspanne: Thesen zur Programmatick und Methodologie.* Paper presented at Entwicklungspsychologie, Augsberg, Germany.

Cohen, E. A., Gelfand, D. M., & Hartmann, D. P. (1981). Causal reasoning as a function of behavioral consequences. *Child Development, 52,* 514-522.

Costanzo, P. R., Grumet, J. F., & Brehm, S. S. (1974). The effects of choice and source of constraint on children's attributions of preference. *Journal of Experimental Social Psychology, 10,* 352-364.

Damon, W. (1971). *A developmental analysis of the positive justice concept from childhood through adolescence.* Unpublished master's thesis, University of California, Berkeley.

DiVitto, B., & McArthur, L.Z. (1978). Developmental differences in the use of distinctiveness, consensus, and consistency information in making causal judgments. *Developmental Psychology, 14,* 474-482.

Dreman, S. B. (1976). Sharing behavior in Israeli school children: Cognitive and social learning factors. *Child Development, 47,* 186-194.

Dreman, S. B., & Greenbaum, C. W. (1973). Altruism or reciprocity: Sharing behavior in Israeli kindergarten children. *Child Development, 44,* 61-68.

Eckensberger, L. H. (1979). A metamethodological evaluation of psychological theories from a cross-cultural perspective. In L. H. Eckensberger, W. J. Lonner, & Y. H. Poortinga (Eds.), *Cross-cultural contributions to psychology.* Lisse: Swets & Zeitlinger.

Eckensberger, L. H. (1981). *The assessment of normative concepts: Some considerations on the affinity between methodological approaches and preferred theories.* Unpublished manuscript, University of Saarland, West Germany.

Eckensberger, L. H., & Reinshagen, H. (1980). Kohlbergs stufentheorie der entwicklung des morabischen urteils: Ein versuch ihrer keinterpretation im bezugrahmen handlungstheorietischer konzepte. In L. H. Eckensberger & R. K. Silbereisen (Eds.), *Entwicklung sozialer kognitionen: Modelle, theorien, methoden, anwendung.* Stuttgart: Klett.

Eckensberger, L. H. & Silbereisen, R. K. (1980a). Einleitung: Handlungstheoretische fur die entwicklungspsycholgie sozialer kognitionen. In L. H. Eckensberger & R. K. Silbereisen (Eds.), *Entwicklung sozialer kognitionen: Modelle, theorien, methoden, anwendung.* Stuttgart: Klett.

Eckensberger, L. H., & Silbereisen, R. K. (1980b). *Entwicklung sozialer kognitionen: Modelle, theorien, methoden, anwendung.* Stuttgart: Klett.

Eisenberg, N. (Ed.). (1982a). *The development of prosocial behavior.* New York: Academic Press.

Eisenberg, N. (1982b). The development of reasoning regarding prosocial behavior. In N. Eisenberg (Ed.), *The development of prosocial behavior.* New York: Academic Press.

Eisenberg, N. (1982c). Introduction. In N. Eisenberg (Ed.), *The development of prosocial behavior.* New York: Academic Press.

Eisenberg, N., Lennon, R., & Roth, L. (1983). Prosocial development: A longitudinal study. *Developmental Psychology, 19,* 846-855.

Eisengberg-Berg, N. (1979a). The relationship of prosocial moral reasoning to altruism, political liberalism, and intelligence. *Developmental Psychology, 15,* 87-89.

Eisenberg-Berg, N. (1979b). Development of children's prosocial moral judgment. *Developmental Psychology, 15,* 128-137.

Eisenberg-Berg, N., & Hand, M. (1979). The relationship of preschoolers' reasoning about prosocial moral conflicts to prosocial behavior. *Child Development, 50,* 356-363.

Eisenberg-Berg, N., & Neal, C. (1979), Children's moral reasoning about their spontaneous prosocial behavior. *Developmental Psychology, 15,* 228-229.

Eisenberg-Berg, N., & Neal, C. (1981). The effects of personality of the protagonist and costs of helping on children's moral judgment. *Personality and Social Psychology Bulletin, 7,* 17-23.

Eisenberg-Berg, N., & Roth, K. (1980). The development of children's prosocial moral judgment: A longitudinal follow-up. *Developmental Psychology, 16,* 375-376.

Gelfand, D. M., & Hartmann, D. P. (1982). Response consequences and attributions: Two contributors to prosocial behavior. In N. Eisenberg (Ed.), *The development of prosocial behavior.* New York: Academic Press.

Goffman, E. (1963) *Behavior in public places.* New York: Free Press.

Goffman, E. (1969). The presentation of self in everyday life. London: Penguin Books.

Guttman, J., Bar-Tal, D., & Leiser, P. (1979). The effect of various reward situations on children's helping behavior. Unpublished manuscript.

Hacker, W. (1978). Allgemeine arbertsund Ingenieurpsychologie. Bern: Huber Verlag.

Heider, R. (1958). The psychology of interpersonal relations. New York: Wiley.

Hoffman, M. L. (1976). Empathy, role-taking, guilt, and development of altruistic motives. In T. Lickona (Ed.), Moral development and behavior: Theory, research and social issues. New York: Holt, Rinehart & Winston.

Hoffman, M. L. (1982). Development of prosocial motivation: Empathy and guilt. In N. Eisenberg (Ed.). The development of prosocial behavior. New York: Academic Press.

Iannotti, R. J. (1978). Effect of role-taking experiences on role taking, empathy, altruism, and aggression. Developmental Psychology, 14, 119-124.

Kalbermaten, V., & von Cranach, M. (in press). Heirarchisch organisierte beobachtungssysteme zur handlungsanalyse. In P. Winkler (Ed.), Method zur analyse von face-to-face situationen. Stuttgart: Metzler.

Karniol, R. (1982). Settings, scripts, and self-schemata: A cognitive analysis of the development of prosocial behavior. In N. Eisenberg (Ed.), The development of prosocial behavior. New York: Academic Press.

Karniol, R., & Ross, M. (1976). The development of causal attributions in social perception. Journal of Personality and Social Psychology, 34, 455-464.

Karniol, R., & Ross, M. (1979). Children's use of a causal attribution schema and the inference of manipulative intentions. Child Development, 50, 464-468.

Kassin, S. M. (1981). From lay child to "layman": Developmental causal attribution. In S. S. Brehm, S. M. Kassin, & F. X. Gibbons (Eds.), Developmental social psychology: Theory and research. New York: Oxford University Press.

Kelley, H. H. (1967). Attribution theory in social psychology. In D. Levine (Ed.) Nebraska Symposium on Motivation: Vol. 15. Lincoln: University of Nebraska Press.

Kelley, H. H. (1972). Attribution in social interaction. In E. E. Jones, D. E. Kanouse, H. H. Kelley, R. E. Nisbett, S. Valines, & B. Weiner (Eds.), Attribution: Perceiving the causes of behavior. Morristown, NJ: General Learning Press.

Kohlberg, L. (1969). Stage and sequence: The cognitive developmental approach to socialization. In D. Goslin (Ed.), The handbook of socialization theory and research. Chicago: Rand McNally.

Kohlberg, L. (1971). From is to ought: How to commit the naturalistic fallacy and get away with it in the study of moral development. In T. Mischel (Ed.), Cognitive development and epistemology. New York: Academic Press.

Krebs, D. L. (1970). Altruism: An examination of the concept and a review of the literature. Psychological Bulletin, 73, 258-303.

Krebs, D. L., & Russel, C. (1981). Role-taking and altruism: When you put yourself in another's shoes, will they take you to the owner's aid? In J. P. Rushton & R. M. Sorrentino (Eds.), Altruism and helping behavior: Social, personality, and developmental perspectives. Hillsdale, NJ: Lawrence Erlbaum.

Kurdek, L. A. (1978). Perspective taking as the cognitive basis of children's moral development: A review of the literature. Merrill-Palmer Quarterly, 24, 3-27.

Langer, E., Blank, A., & Chanowitz, B. (1978). The mindlessness of ostensibly thoughtful action: The role of "placebic" information in interpersonal attraction. Journal of Personality and Social Psychology, 36, 635-642.

42   *Prosocial Processes*

Lantermann, E. D. (1980). *Interaktionen: Person, situation und handlung.* Munden: Urban & Schwarzenberg.

Leahy, R. L. (1979). Development of conceptions of prosocial behavior: Information affecting rewards given for altruism and kindness. *Developmental Psychology, 15,* 34-37.

Miller, G. A., Galanter, E., & Pribram, K. H. (1960). *Plans and the structure of behavior.* New York: Holt.

Mussen, P., & Eisenberg-Berg, N. (1977). *Roots of caring, sharing, and helping: The development of prosocial behavior in children.* San Francisco: Freeman.

Pearl, R. A. (1979, March). *Developmental and situational influences on children's understanding of prosocial behavior.* Paper presented at the Biennial Meeting of the Society for Research in Child Development, San Francisco.

Peterson, L. (1980). Developmental changes in verbal and behavioral sensitivity to cues of social norms of altruism. *Child Development, 51,* 830-838.

Piaget, J. (1965). *The moral judgment of the child.* New York: Free Press. (Original work published 1932)

Raviv, A., Bar-Tal, D., & Lewis-Levin, T. (1980). Motivations for donation behavior by boys of three different ages. *Child Development, 51,* 610-613.

Ruble, D. N., & Rholes, W. S. (1981). The development of children's perceptions and attributions about their social world. In J. H. Harvey, W. J. Ickes, & R. F. Kidd (Eds.), *New directions in attributional research: Vol. 3.* Hillsdale, NJ: Lawrence Erlbaum

Rushton, J. P. (1975). Generosity in children: Immediate and long-term effects of modeling, preaching, and moral judgment. *Journal of Personality and Social Psychology, 31,* 459-466.

Rushton, J. P. *Altruism, socialization, and society.* Englewood Cliffs, NJ: Prentice-Hall.

Shure, M. B. (1968). Fairness, generosity, and selfishness: The naive psychology of children and young adults. *Child Development, 30,* 857-886.

Smith, C. L., Gelfand, D. M., Hartmann, D. P., & Partlow, M. P. (1979). Children's causal attributions regarding help giving. *Child Development, 50,* 203-210.

Staub, E. (1979). *Positive social behavior and morality: Socialization and development: Vol. 2.* New York: Academic Press.

Suls, J., Witenberg, S., & Gutkin, D. (1981). Evaluating reciprocal and nonreciprocal prosocial behavior: Developmental changes. *Personality and Social Psychology Bulletin, 7,* 25-31.

Ugurel-Semin, R. (1952). Moral behavior and moral judgment of children. *Journal of Abnormal and Social Psychology, 47,* 464-474.

Underwood, B., & Moore, B. (1982). Perspective-taking and altruism. *Psychological Bulletin, 91,* 143-173.

Volpert, W. (1974). *Handlungsstrukturanalyse als Beitrag zur Qualifikationsforschung.* Koln: Paul Rugenstein.

von Cranach, M. (in press). The organisation of goal-directed action: A research report. In M. Brenner & M. von Cranach (Eds.), *Discovery strategies in the psychology of action.* Cambridge, England: Cambridge University Press.

von Cranach, M., & Harre, R. (1982). *The analysis of action: Recent theoretical and empirical advances.* New York: Cambridge University Press.

von Cranach, M., Kalbermatten, U., Indermuhle, K., & Gugler, B. (1982). *Goal-directed action.* New York: Academic Press.

2

Perspective Taking and Empathy in the Child's Prosocial Behavior

Mark A. Barnett

While the study of nonsocial cognition has been one of the most extensively explored areas in developmental psychology, investigation of the child's growing awareness of the thoughts, needs, intentions, and emotions of other individuals is in its infancy. One assumption that has presumably spurred interest in children's social cognition, but that has received only limited empirical attention, is that the manner in which children conceptualize other individuals is related to the manner in which they interact with those individuals (Shantz, 1975). The primary purpose of this chapter is to examine the evidence relevant to a portion of this assumption. More specifically, the present discussion will focus on two related aspects of the child's evolving social awareness— perspective taking and empathy—and the role they play in the development and expression of prosocial behavior.

THEORETICAL VIEWS ON PERSPECTIVE TAKING
AND PROSOCIAL BEHAVIOR

The child's ability to take the role, or perspective, of other individuals has frequently been viewed as crucial to the development
and maintenance of effective interpersonal relations. While the empirical study of the relationship between perspective-taking
abilities and various moral indices in children and adults is a
relatively recent development, the belief that one's capacity or inclination to assume the role of another is central to moral thought
and behavior is certainly not a new one. At the turn of the century, E. A. Ross wrote (with apparent conviction and indignation):

> In every cluster there are predatory persons—moral idiots or moral
> lunatics, who can no more put themselves in the place of another
> than the beast can enter into the anguish of its prey or the parasite
> sympathize with his host [cited in Maccoby, 1980, p. 150].

Other writers (for example, Mead, 1934; Weinstein, 1969), generally less descriptive in their formulations, have also attributed major importance to the individual's ability to consider another's viewpoint in the course of establishing and securing adaptive social
relations.

Much of the credit for giving impetus to the systematic study
of perspective taking and interpersonal behaviors belongs to the
Swiss psychologist Jean Piaget (1926, 1932/1965, 1967; Piaget
& Inhelder, 1969). On the basis of his observations of and conversations with children of various ages, Piaget concluded that prior
to the age of six, children are egocentric and unable to differentiate between their own and others' perceptions, thoughts, and
emotional states. He suggested that young children's inability to
"decenter," or adequately consider the point of view of another,
caused them to be ineffective communicators and relatively immature in moral judgment and behavior.

In a similar vein, Piaget proposed that the young, "self-centered"
child cannot engage in prosocial and cooperative acts that require
the comprehension of others' needs. According to his theory, the
decrease in egocentrism, and the concomitant increase in the ability to decenter, both of which occur at about seven or eight years
of age, are the result of the child's advancement in general
cognitive skills, and of more frequent interaction and confronta-
tion with peers who differ in their wishes, perspectives, needs, and

thoughts. With regard to the latter point, just as peer interaction is seen as necessary for the development of role-taking skills, so the enhancement of role-taking abilities is said to enable the child to engage in more mature forms of positive social interaction. Piaget believed, therefore, that during middle childhood the enactment of positive and reciprocal social behaviors, such as effective communication, cooperation, helping, and sharing, serves to strengthen and is strengthened by the ability to assume the perspective of another.

Several contemporary theorists and researchers (for example, Feffer, 1970; Flavell, 1974; Selman, 1976; Selman & Byrne, 1974), expanding on Piaget's ideas concerning the child's social cognition and social behavior, have offered various "stage" models to describe the development of perspective-taking and related abilities. While the individual models differ on many dimensions, there appears to be some consistency in viewing children as progressing from (1) an egocentric view of their social world, to (2) understanding that other individuals may have contrasting perceptions, thoughts, and feelings, to (3) consideration of one's own and another's point of view in a separate and successive manner, to (4) consideration of a number of different perspectives in a coordinated, balanced, and simultaneous manner. In contemporary theorists' process of examining, clarifying, and extending Piaget's notions, however, a basic assumption remains: The shift away from an early egocentric orientation underlies the child's improvement in interpersonal skills and the development of moral standards and behaviors that reflect concern for the feelings and welfare of others (Flavell, Botkin, Fry, Wright, & Jarvis, 1968).

RESEARCH ON PERSPECTIVE TAKING AND PROSOCIAL BEHAVIOR

This section will examine the research evidence relevant to two assumptions derived from the theoretical views on perspective taking and its role in the child's interpersonal behavior. The first assumption is that there is considerable overlap, or generality, among the various "types" of perspective-taking abilities. According to Staub (1979, p. 75):

Generality has been expected on the basis of the conception advanced by Piaget that egocentrism is an inability to differentiate be-

tween one's view and that of others that is not tied to any particular
"content" area.

Therefore, a child's scores on various measures of his or her ability to infer the perceptions, thoughts, and feelings of other individuals would be expected to be highly associated, since each
type of perspective taking is merely a portion of a larger social
cognitive capacity.

The second assumption is central to the present discussion and
concerns the relationship between a child's perspective-taking
capacity and his or her prosocial behavior. Piaget's bidirectional
causal relation notion, outlined earlier, suggests that peer interaction is necessary for the development of perspective-taking skills,
and that the development of perspective-taking skills is instrumental in upgrading the quality of a child's peer interactions. Thus,
differences in children's perspective-taking abilities would be expected to parallel differences in the extent to which they engage
in particular prosocial behaviors.

To a large extent, the degree to which these assumptions can
be adequately tested is a function of the adequacy of the in-
struments designed to assess the various perspective-taking competencies. Prior to investigating these two assumptions, a brief
overview of some common approaches to the measurement of
perspective-taking skills will be presented.

Measurement of Perspective-Taking Ability

While numerous tasks have been developed to assess various
aspects of children's social inferential ability, perspective-taking
measures can be grouped into three general categories: perceptual, cognitive, and affective (Shantz, 1975). Each of these
classifications is briefly described below. Within each category,
the procedure involved in one or two tasks is presented as merely
illustrative of the general approach to measurement within that
domain. (For a more complete discussion of the assessment of
perspective-taking abilities, see the excellent reviews by Shantz,
1975, and Ford, 1979.)

Perceptual perspective-taking tasks (see for example, Flavell et
al., 1968; Kurdek, 1980) have typically been designed to assess
a child's ability to infer what is seen by an individual viewing an
object or array of objects from a different physical perspective.
In Piaget and Inhelder's (1956) "three mountains" problem, for ex-

ample, a child is seated at a table on which three different-sized mountains are displayed. A doll is placed in one of the three vacant chairs positioned around the table, and the child is asked to identify how the doll "sees" the mountain landscape by selecting one of several drawings or photographs, by drawing a picture of what the doll sees, or by reconstructing the doll's perspective with a set of materials provided to the child. If the child disregards the doll's location and attributes his or her own visual perspective to it, then the child's response is scored as egocentric. On the other hand, a correct response is interpreted as evidence of perceptual, or spatial, perspective-taking ability. Modifications of the original three mountains task have generally utilized stimulus arrays or materials that vary in their familiarity and complexity.

Cognitive perspective-taking tasks (see for example, Chandler, 1973; Feffer, 1970; Flavell et al., 1968; Glucksberg, Krauss, & Higgins, 1975; Selman, 1971) assess a child's ability to infer the thoughts, motives, attitudes, or intentions of another individual. This category of perspective taking is extremely broad and, consequently, many diverse research strategies have been developed to measure the child's capacity "to think about what the other is thinking about."

Studies of children's competence in communicating (for example, Applegate & Delia, 1980; Delia & Clark, 1977; Glucksberg et al., 1975) have generally attempted to assess the extent to which children use and adapt their language to inform or persuade another person. In Chandler's (1973) "bystander task," a child is first asked to describe a story depicted in a multiframe cartoon sequence and then to retell the story from the perspective of a bystander who is introduced midway through the sequence and who is presumably unaware of the preceding events. The child's perspective-taking score on this task is determined by the extent to which his or her second narrative is restricted to the limited perspective of the bystander. Feffer's Role-Taking Task (Feffer, 1959, 1970) is conceptually similar to Chandler's and requires a child to tell and retell a story as it would be perceived or experienced by each of several characters depicted in a picture. Other measurement approaches within this domain (see Ford, 1979) tap children's public and private speech in naturalistic settings, their understanding of the recursive nature of thought, and their ability to infer the strategy of an opponent in a game.

Affective perspective-taking tasks (see, for example, Borke, 1971; Rothenberg, 1970) assess the ability to perceive, identify, and accurately infer the emotional state of another individual. In Borke's (1971, 1973) widely used Interpersonal Perception Test, a child is shown a series of pictures with accompanying narrations that depict a boy or girl in affect-laden situations, such as losing a pet (sadness) or being lost (fear). The main character's face is left blank in each story picture, and the child's task is to identify correctly the picture of the facial expression that corresponds to the way the story character felt.

Several researchers (for example, Burns & Cavey, 1957; Deutsch, 1974; Gnepp, Klayman, & Trabasso, 1982; Greenspan, Barenboim, & Chandler, 1976) have presented children with pictures or films in which the facial expression (for example, a frown) of the central story character is displayed but not always congruent with the situational cues (for example, a birthday party). Of particular interest here is the determination of the extent to which the child's attribution of another's affect is influenced by the facial and contextual cues. Finally, Rothenberg's (1970) affective perspective-taking measure requires a child to describe, as well as explain, the feelings of adult characters presented in a series of vignettes.

Relationship Among Indices of Perspective-Taking Ability

As outlined earlier, the initial assumption to be examined is that egocentrism and decentering are general and unitary constructs and that a child's scores on various indices of perspective-taking ability should thus be closely associated. Stated simply, the empirical support for this contention is quite weak. Prior reviews of the perspective-taking literature (Ford, 1979; Krebs & Russell, 1981; Kurdek, 1978a; Rubin, 1973, 1978; Shantz, 1975) have indicated that the majority of the intercorrelations reported in individual studies fall in the nonsignificant to low-but-significant range, suggesting that "there is, at best, only a moderate relationship among various role-taking skills" (Shantz, 1975, p. 43). Surprisingly, the relationship among scores on different tasks within a particular category of perspective-taking ability is typically neither stronger nor more consistent than the relationship among scores on tasks from different categories. For example, Knudson and Kagan (1977, p. 243) administered seven perceptual perspective-taking tasks to five- and nine-year-old children and concluded that the "low and inconsistent correlations among the visual perspective role-taking tasks indicate that visual perspective role-taking is not an established unidimensional construct."

Several methodological problems have been suggested as contributing to the pattern of weak intercorrelations among the various indices of children's perspective-taking ability. Among the measurement issues commonly noted are the marked differences across tasks in the complexity and familiarity of the stimuli used, the level of role taking required, and the mode of response (for example, verbal versus nonverbal) required.[1] However, even in a study in which perspective-taking tasks were specifically designed to minimize the variation in stimuli, format, and response factors (Hudson, 1978), little relationship was found among second graders' scores on the measures. In sum, the weight of the evidence fails to support the notion that egocentrism and decentering are unitary constructs: What the data do indicate is that perspective taking is a complex and multidimensional social cognitive skill.

Relationship Between Measures of Perspective Taking and Prosocial Behavior

The second assumption to be investigated is that social cognition and social behavior are highly interrelated and, more specifically, that differences in children's perspective-taking abilities are associated with differences in the degree to which they perform prosocial acts. Correlational studies designed to address this issue have typically assessed the relationship between children's scores on one or more perspective-taking measures and their scores on one or more prosocial behavior indices, such as the number of tokens or candies shared with a friend or needy other, or the extent of helping and comforting as determined by naturalistic observation.[2]

The results of these studies have been highly inconsistent. While some studies (Buckley, Siegel, & Ness, 1979; Rubin & Schneider, 1973) report positive associations reaching levels of statistical significance, others (Abroms & Gollin, 1980; Eisenberg-Berg & Lennon, 1980) report children's scores on the measures to be unrelated. To add to the confusion, several studies (Hudson, Forman, & Brion-Meisels, 1982; Iannotti, 1978) that have utilized multiple measures of perspective taking and/or positive social behavior inexplicably found a mixture of unrelated, significantly positively related, and in some cases significantly negatively related pairs of indices. Moreover, as Kurdek's (1978a) review indicates, no clear pattern emerges even when particular measures or categories of measures are considered.[3]

How does one account for what Krebs (1980, p.1) has described as "a perfectly obvious relationship" and "a set of entirely inconsistent conclusions"? Certainly, there are measurement problems to consider. Since children's scores on the various perspective-taking measures do not "stick together" very well, it is unlikely that a study assessing the association between children's scores on several social inference measures and even a single index of prosocial behavior would yield a consistent pattern of results. Moreover, helping, like perspective taking, is not a unitary construct. Therefore, studies that use multiple measures of both perspective-taking competence and prosocial behavior would seem to have little chance of achieving clear and robust findings.

In a related vein, the prosocial measures used in correlational studies have differed markedly in the extent to which they are likely to tap or require the perspective-taking skills being assessed. Even in those cases in which perspective-taking and prosocial measures appear to be conceptually related, the studies have failed to determine whether the children who are identified as "perspective takers" are actually assuming the role of the person they are given the opportunity to assist (Krebs & Russell, 1981). By assessing only the extent to which children *can* infer the perceptions, thoughts, and feelings of others and not the extent to which they spontaneously do so when confronted with a needy other, these studies may have provided extremely conservative tests of this social cognition-social behavior relationship.

Another more basic flaw may have contributed to the inconsistent findings. Knowledge of the thoughts and feelings of a needy other is likely insufficient, in and of itself, to motivate an individual to engage in a prosocial act. While a child's awareness of another's sadness or distress may arouse, under some circumstances, social norms that impel a helpful or generous response, the child's perspective-taking skills may also be utilized in a manipulative, deceptive, hurtful, and self-serving manner. Germane to the latter point is a brief anecdote described by Staub (1979, p. 72):

> A 2-year-old, having been frustrated and angered by an age-mate, picked up the other child's teddy bear, a favorite toy and security blanket, and threw it out the window. An understanding of the other's probable emotional reaction was presumably present.

The "darker side of perspective taking" is also seen in the results of a correlational study by Kurdek (1978b), involving somewhat older children. In this investigation, first- through fourth-grade children who were classified as good perspective takers on a battery of traditional role-taking measures were rated by their teachers, not as helpful and cooperative, but as discipline problems, highly disruptive, and prone to fighting and quarreling. Apparently, the availability of particular perspective-taking skills does not ensure that a child will act (or will act specifically in an other-oriented manner) when confronted with the needs and vulnerabilities of others. Thus, the perspective taking-prosocial behavior relation may be a weak and inconsistent one because it lacks a crucial motivational component. Such a component may be found in the construct of empathy.

EMPATHY AND PROSOCIAL BEHAVIOR

Children's capacity to empathize with or vicariously experience the emotional state of another individual has frequently been suggested as serving an important role in the enactment of prosocial behavior (Barnett, 1982; Mussen & Eisenberg-Berg, 1977; Staub, 1978). The empathically aroused individual is commonly thought to be motivated to aid a needy other by the anticipated cessation of the mutually experienced distress and/or the anticipated vicarious pleasure following the helpful act. Although certain social cognitive abilities may be necessary for the elicitation of an empathic response (a point that will be discussed in greater detail later), it is the observer's emotional response, rather than the mere cognitive awareness of the other's plight or affective state, that is believed to be instrumental in impelling subsequent prosocial acts.

Correlational Studies:
Measurement Problems and Issues

Studies of the relationship between the young child's empathic tendency and prosocial behavior have frequently utilized the Feshbach Affective Situations Test of Empathy (FASTE; Feshbach & Roe, 1968) as the measure of empathy. This measure, which bears some resemblance to Borke's (1971, 1973) affective perspective-taking task, consists of four pairs of slide sequences in which young children are shown in situations designed to elicit the emotions of happiness, sadness, anger, and fear. A brief narration describing the events in the slides accompanies each sequence. After viewing a complete slide sequence, the child is asked, "How do you feel?" and the responses are recorded for later analysis. Each response is rated on the following 3-point scale for the degree to which it matches the affect of the child featured in a slide sequence: 0 = incorrect response; 1 = generally correct emotion on the positive-negative affect dimension; 2 = correct, specific emotion. The total empathy score for each child is thus the sum for all four emotions (range = 0-16).

Although some positive findings have been reported (Feshbach, 1973; Marcus, Telleen, & Roke, 1979), the majority of studies have found that scores on the FASTE (or modified versions of it) are unrelated to various prosocial indices (for example, Eisenberg-Berg & Lennon, 1980; Sawin, 1979). One factor that may have served to attenuate the dispositional empathy-prosocial behavior relation in some studies is the selection of a prosocial index, such as cooperation on a game (Levine & Hoffman, 1975), which may be relatively devoid of empathy-arousing cues. In addition, the measure of empathy may be suspect. Concerning the latter point, several attacks (Eisenberg-Berg & Lennon, 1980; Sawin, 1979) have recently been leveled against the Affective Situations Test, including criticisms of the measure's (1) emphasis on the correctness, rather than the intensity, of the empathic response; (2) verbal bias, which may yield scores confounded with the child's cognitive ability, age, or sex; (3) assumption that a child's "emotional bond" to a story character develops rapidly and is easily manipulated; and (4) vulnerability to demand characteristics and social desirability. In addition, since a child's tendency to empathize with positive and negative affects in others may be only weakly related (Howard, 1981; Marcus, 1980) and differentially related to the enactment of prosocial behaviors (Feshbach, 1980), the exclusive use of "total" empathy scores in many prior studies may have served to obscure any positive findings.

Alternative measures of empathy in children have recently been proposed. Barnett, Howard, Melton, and Dino (1982) found teacher and average peer ratings of sixth graders' empathic dispositions to be significantly correlated in each of seven classes (r's ranging from .51 to .80; overall r = .54, p < .001). In this study, a composite index of teacher and peer ratings was used to establish high- and low-empathy groups. Bryant's (1982) paper-and-pencil measure of empathy for children and adolescents was adapted from the Mehrabian and Epstein (1972) adult empathy scale. It has demonstrated "satisfactory reliability and preliminary construct validity" (Bryant, 1982, p. 413).

Nonverbal indices of emotional arousal, such as facial expression and physiological response, also show potential as measures of dispositional empathy in children. Concerning the former type of response, marked individual differences have been found in the extent to which children spontaneously match their own facial expressions to those of sad characters depicted in affect-laden slide or film presentations (Buck, 1975; Hamilton, 1973; Leiman, 1978). Moreover, Sawin (1979) reports that an empathy index based on ratings of first graders' facial expressions and tone of voice in response to the Feshbach and Roe (1968) slides was a better predictor of generosity (sharing prize marbles with a less fortunate child) than was the conventional empathy score based on self-reported affect.

Physiological indices of empathic arousal, which have been found to be associated with helping in adults (Krebs, 1975), might also be used effectively in studies with young children. However, it is extremely difficult to obtain a valid and reliable physiological measure of a specific emotion at any age (see review by Yarrow, 1979). For example, when some children (or adults) are confronted with the distress of another individual, particularly in a strange laboratory, changes in electrodermal response or heart rate may reflect a startle response or fear rather than vicarious affect arousal. Moreover, large differences have been found in the extent to which individuals are physiologically labile. Indeed, Buck (1975) has distinguished between groups of children (as well as adults) who are physiologically responsive ("internalizers") and those who tend to display emotions facially ("externalizers").

The developing child's increasing awareness and use of "display rules" (Ekman, Friesen, & Ellsworth, 1972; Yarczower & Daruns, 1982), which can function to suppress or distort the overt display of certain emotions in certain situations, add to the difficulty of identifying a single measure of empathy that is appropriate for use throughout childhood. Perhaps the use of multiple measures of empathy in future studies, in addition to providing insight into individual differences in modes of response, will enable researchers to predict a child's prosocial behavior better than any single index (Barnett, 1982; Bryant, 1982; Sawin, 1979).

Experimental and Naturalistic Studies

Experimental studies of the empathy/helping relation have typically investigated the effects of directing a child's attention to an unfortunate other's distress. One weakness of much of the experimental research conducted to date is that empathy has generally been assumed to be aroused rather than directly measured. Nonetheless, the results of recent investigations (Eisenberg-Berg & Geisheker, 1979; Howard & Barnett, 1981) indicate that encouraging children to focus on and share the feelings of needy others elicits a more charitable response than an appeal that encourages them to think about less fortunate others but that makes no mention of feelings.

In an investigation in which an "in task" assessment of empathy was utilized, Leiman (1978) found that five- and six-year-old children who were judged as displaying empathic facial expressions upon seeing a same-sex actor lose a favored marble collection subsequently worked harder on a task (the "marble donation machine") to replace the actor's play materials than did children whose facial expressions had been rated as neutral and non-empathic. Other investigations (Barnett, Howard, King, & Dino, 1981; Barnett, King, & Howard, 1979) have indicated that focusing an individual's attention on another's misfortune and sadness may not only enhance the tendency to aid the person who was the original source of concern, but may promote a generalized inclination to assist others. However, Barnett et al. (1982) found that briefly inducing sadness about an unfortunate playmate heightened the subsequent charity of high, but not low, empathic children to a group of sick children in a hospital. Thus it appears that even momentarily directing the highly empathic child's attention to another's misfortune may be sufficient to make the feelings and needs of others generally more salient and thereby enhance the expression of prosocial behaviors. For the less empathic child, on the other hand, the activation of helping norms, perhaps coupled with more intensive and involving procedures for inducing empathy, may be necessary to elicit concerns relevant to helping.

Research conducted in more naturalistic settings (Bridges, 1931; Eisenberg-Berg & Neal, 1979; Murphy, 1937) has indicated that many young children tend to respond to another child's misfortune by staring with an anxious expression or by displaying other nonverbal signs of distress. Furthermore, Murphy (1937) observed that when an emotionally responsive child helped, his or her affective response typically diminished; however, when such a child failed to help, the nonverbal cues of distress were generally prolonged.

Although young children typically display some response when observing distress in others, they appear to differ markedly in their tendency or willingness to respond in a sensitive and helpful way. In this regard, Zahn-Waxler and Radke-Yarrow (1979, 1982; Radke-Yarrow & Zahn-Waxler, in press) have reported stable and patterned individual differences in empathic responses among one- and two-year-olds, as well as individual continuity to age seven in children's intensity, complexity, and mode of response to others' emotions. Whereas the prosocial interactions of some children in their investigations were found to have an intense affective component, other children tended to react to another's distress in an unemotional manner (for example, inspecting, exploring, asking questions), in an aggressive manner (hitting a person who made a baby cry), or in an anxious and "guilty" manner (turning and running away). Assessing the antecedents of such early individual differences in future studies will serve to expand our understanding of empathy and its role in the development and expression of prosocial behavior.

PERSPECTIVE TAKING, EMPATHY, AND PROSOCIAL BEHAVIOR: INTEGRATION AND CONCLUSION

To this point, perspective taking and empathy have been treated as largely separate and distinct constructs, each occupying a unique role in children's prosocial behavior. In reality, the child's cognitive awareness of and affective responsiveness to the distress of a needy other are complexly intertwined and frequently act together in eliciting a helpful response.

Recent models of empathy and positive social behavior have emphasized the development of the requisite cognitive skills for experiencing and properly interpreting vicarious affective arousal. According to Feshbach's (1978) three-component model, for example, an empathic response requires (1) the ability to discriminate and identify the emotional states of other individuals, (2) the capacity to take the perspective or role of the other, and (3) the evocation of a shared affective response. To Feshbach, the cognitive and affective components relevant to the expression of empathy are highly interrelated and are essential ingredients in a child's enactment of positive social behavior. Consistent with this notion are the results of training studies with children (Chandler, 1973; Feshbach, 1979; Iannotti, 1978; Spivack & Shure, 1974; Staub, 1971) which demonstrate that structured experiences promoting role taking and sensitivity to the feelings of others can serve to heighten the expression of prosocial behavior and decrease the incidence of antisocial behavior.

Hoffman's (1975a, 1977a) developmental model is an ambitious attempt to explain how cognitive and affective factors play changing and interactive roles in children's evolving altruistic behavior. From this perspective, the child's emerging ability to comprehend the distinction between self and other, and the growing awareness that other individuals have internal states and feelings separate from one's own, set the stage for higher levels of empathic responding. Hoffman suggests that although infants demonstrate an affective orientation to others that may represent a "constitutionally based, early precursor of empathy" (Hoffman, 1977b, p. 299), role-taking processes interact with the individual's empathic "disposition" throughout childhood to engender the motivation to behave altruistically. Thus, while young children frequently respond to obvious displays of distress by others, the circumstances that will elicit emotional arousal and a helpful response are believed to change and broaden with the child's increasing experiences and cognitive growth.

The complex interaction of cognitive and affective factors is also seen in the antecedents of altruistic behavior in children. Parental affection appears to play a particularly important role in early expressions of prosocial behavior, presumably by satisfying children's own emotional needs and thereby enabling them to be aware of and emotionally receptive to the needs of others (Hoffman, 1975b). In addition, generosity and consideration for others have been found to be enhanced by inductive socialization techniques that encourage children to imagine themselves in another's place and that simultaneously direct their attention to the potential impact of their behavior on the feelings of others (Perry, Bussey, & Frieberg, 1981). In a similar vein, Radke-Yarrow and Zahn-Waxler (in press) have concluded from their extensive observations that heightened prosocial behavior in young children is associated with mothers who frequently (1) convey a clear cognitive message to their child (an explanation or demonstration) of the consequences of his or her behavior for the victim, and (2) reinforce this message with a display of intense emotion and statements of their principles and expectations for the child's behavior. Thus it appears that in the caregiver's response to the child, as in many other aspects of prosocial behavior, "the compound of cognitive and affective components is essential" (Radke-Yarrow & Zahn-Waxler, in press).[4]

Finally, while social inferential skills and emotional responsiveness play important and reciprocal roles in the development and expression of prosocial behavior, there are undoubtedly circumstances under which they are insufficient to impel a generous or caring response. For example, children who experience distress and concern when confronted with an unfortunate other may fail to offer assistance if they perceive little personal responsibility to help, or if they feel unable to help or comfort effectively (Burleson, 1982). On some occasions the child may become so overwhelmed by the distress of another individual that egoistic concerns predominate over a concern for the other, producing avoidance or even derogation of the victim. In addition, children who have had limited helping experience may not have learned that effective intervention can function to alleviate the other's, as well as one's own, discomfort. These potential "failures to help" suggest that during the course of their social-emotional development, children need more than just expanding capacities to infer and experience the feelings of others; they need to be provided with the encouragement, the opportunities, and the interpersonal skills necessary to translate their thoughts and feelings into appropriate prosocial behavior.

NOTES

1. For a discussion of other psychometric properties (for example, reliability, internal consistency) of various perspective-taking tasks, see Ford (1979), Kurdek (1978a), Shantz (1975), and Rubin (1978).

2. Although the present review is largely restricted to a consideration of the helping literature, it should be noted that research has also been conducted on the association between children's perspective-taking ability and their popularity (Krantz, 1982; McGuire & Weisz, 1982; Rothenberg, 1970; Rubin, 1972, 1973), social competence (Rubin, 1982), interpersonal problem solving (Marsh, Serafica, & Barenboim, 1981; Spivack, Platt, & Shure, 1976), teacher-rated classroom behavior (Burka & Glenwick, 1978; Kurdek, 1978b), moral judgment (Krebs & Gillmore, 1982; Kurdek, 1978a, 1980), and nonverbal decoding skills (White & Feldman, 1981).

3. Underwood and Moore's (1982) recent meta-analysis of the research findings generally presents a more favorable view of the perspective taking-prosocial

behavior relation than that presented here or in prior reviews. Nonetheless, the authors did conclude that "even where reliable associations between altruism and perspective-taking measures have been identified, the relationships are not dramatically large" and may frequently be mediated by other variables (Underwood & Moore, 1982, p. 169).

4. For a general discussion of the role of caregiver communication in children's social cognitive development, see Applegate and Delia (1980).

REFERENCES

Abroms, K. I., & Gollin, J. B. (1980). Developmental study of gifted preschool children and measures of psychosocial giftedness. *Exceptional Children, 46,* 334-341.

Applegate, J. L., & Delia, J. G. (1980). Person-centered speech, psychological development, and the contexts of language usage. In R. St. Clair & H. Giles (Eds.), *Social and psychological contexts of language.* Hillsdale, NJ: Lawrence Erlbaum.

Barnett, M. A. (1982). Empathy and prosocial behavior in children. In T. M. Field, A. Huston, H. C. Quay, L. Troll, & G. E. Finley (Eds.), *Review of human development.* New York: Wiley.

Barnett, M. A., Howard, J. A., King, L. M., & Dino, G. A. (1981). Helping behavior and the transfer of empathy. *Journal of Social Psychology, 115,* 125-132.

Barnett, M. A., Howard, J. A., Melton, E. M., & Dino, G. A. (1982). Effect of inducing sadness about self or other on helping behavior in high and low empathic children. *Child Development, 53,* 920-923.

Barnett, M. A., King, L. M., & Howard, J. A. (1979). Inducing affect about self or other: Effects on generosity in children. *Developmental Psychology, 15,* 164-167.

Borke, H. (1971). Interpersonal perception of young children: Egocentrism or empathy? *Developmental Psychology, 5,* 263-269.

Borke, H. (1973). The development of empathy in Chinese and American children between three and six years of age: A cross-cultural study. *Developmental Psychology, 9,* 102-108.

Bridges, K.M.B. (1931). *The social and emotional development of the preschool child.* London: Kegan Paul.

Bryant, B. K. (1982). An index of empathy for children and adolescents. *Child Development, 53,* 413-425.

Buck, R. (1975). Nonverbal communication of affect in children. *Journal of Personality and Social Psychology, 31,* 644-653.

Buckley, N., Siegel, L. S., & Ness, S. (1979). Egocentrism, empathy, and altruistic behavior in young children. *Developmental Psychology, 15,* 329-330.

Burka, A. A., & Glenwick, D. S. (1978). Egocentrism and classroom adjustment. *Journal of Abnormal Child Psychology, 6,* 61-70.

Burleson, B. R. (1982). The development of comforting communication skills in childhood and adolescence. *Child Development, 53,* 1578-1588.

Burns, N., & Cavey, L. (1957). Age differences in empathic ability among children. *Canadian Journal of Psychology, 11,* 227-230.

Chandler, M. J. (1973). Egocentrism and antisocial behavior: The assessment and training of social perspective-taking skills. *Developmental Psychology, 9,* 326-332.

Delia, J. G., & Clark, R. A. (1977). Cognitive complexity, social perception, and the development of listener-adapted communication in six-, eight-, ten-, and twelve-year-old boys. *Communication Monographs, 44,* 326-345.

Deutsch, F. (1974). Female preschoolers' perceptions of affective responses and interpersonal behavior in videotaped episodes. *Developmental Psychology, 10,* 733-740.

Eisenberg-Berg, N., & Geisheker, E. (1979). Content of preachings and power of the model/preacher: The effect on children's generosity. *Developmental Psychology, 15,* 168-175.

Eisenberg-Berg, N., & Lennon, R. (1980). Altruism and the assessment of empathy in the preschool years. *Child Development, 51,* 552-557

Eisenberg-Berg, N., & Neal, C. (1979). Children's moral reasoning about their own spontaneous prosocial behavior. *Developmental Psychology, 15,* 228-229.

Ekman, P., Friesen, W., & Ellsworth, P. (1972). *Emotion in the human face.* New York: Pergamon Press.

Feffer, M. (1959). The cognitive implications of role-taking behavior. *Journal of Personality, 27,* 152-168.

Feffer, M. (1970). Developmental analysis of interpersonal behavior. *Psychological Review, 77,* 197-214.

Feshbach, N. D. (1973, August). *Empathy: An interpersonal process.* Paper presented at the meeting of the American Psychological Association, Montreal.

Feshbach, N. D. (1978). Studies of empathic behavior in children. In B. A. Maher (Ed.), *Progress in experimental personality research: Vol. 8.* New York: Academic Press.

Feshbach, N. D. (1979). Empathy training: A field study in affective education. In S. Feshbach & A. Fraczek (Eds.), *Aggression and behavior change: Biological and social processes.* New York: Praeger.

Feshbach, N. D. (1980, May). *The psychology of empathy and the empathy of psychology.* Paper presented at the meeting of the Western Psychological Association, Honolulu, Hawaii.

Feshbach, N. D., & Roe, K. (1968). Empathy in six- and seven-year-olds. *Child Development, 39,* 133-145.

Flavell, J. H. (1974). The development of inferences about others. In T. Mischel (Ed.), *Understanding other persons.* Oxford: Blackwell, Basil, Mott.

Flavell, J. H., Botkin, P. T., Fry, C. L., Wright, J. W., & Jarvis, P. E. (1968). *The development of role-taking and communication skills in children.* New York: Wiley.

Ford, M. E. (1979). The construct validity of egocentrism. *Psychological Bulletin, 86,* 1169-1188.

Glucksberg, S., Krauss, R. M., & Higgins, R. (1975). The development of communication skills in children. In F. Horowitz (Ed.), *Review of child development research: Vol. 4.* Chicago: University of Chicago Press.

Gnepp, J., Klayman, J., & Trabasso, T. (1982). A hierarchy of information sources for inferring emotional reactions. *Journal of Experimental Child Psychology, 33,* 111-123.

Greenspan, S., Barenboim, C., & Chandler, M. J. (1976). Empathy and pseudo-empathy: The affective judgments of first- and third-graders. *Journal of Genetic Psychology, 129,* 77-88.

Hamilton, M.L. (1973). Imitative behavior and expressive ability in facial expression of emotion. *Developmental Psychology, 8,* 138.

Hoffman, M. L. (1975a). Developmental synthesis of affect and cognition and its implications for altruistic motivation. *Developmental Psychology, 11,* 607-622.

Hoffman, M. L. (1975b). Altruistic behavior and the parent-child relationship. *Journal of Personality and Social Psychology, 31,* 937-943.

Hoffman, M. L. (1977a). Empathy, its development and prosocial implications. In C. B. Keasey (Ed.), *Nebraska symposium on motivation: Vol 25.* Lincoln: University of Nebraska Press.

Hoffman, M. L. (1977b). Personality and social development. *Annual Review of Psychology, 28,* 295-321.

Howard, J. A. (1981). *Preschoolers' empathy for specific affects and their social interaction.* Unpublished manuscript, Kansas State University.

Howard, J. A., & Barnett, M. A. (1981). Arousal of empathy and subsequent generosity in young children. *Journal of Genetic Psychology, 138,* 307-308.

Hudson, L. M. (1978). On the coherence of role-taking abilities: An alternative to correlational analysis. *Child Development, 49,* 223-227.

Hudson, L. M., Forman, E. A., & Brion-Meisels, S. (1982). Role taking as a predictor of prosocial behavior in cross-age tutors. *Child Development, 53,* 1320-1329.

Iannotti, R. J. (1978). The effect of role-taking experiences on role taking, empathy, altruism, and aggression. *Developmental Psychology, 14,* 119-124.

Knudson, K.H.M., & Kagan, S. (1977). Visual perspective role-taking and field-independence among Anglo American and Mexican American children of two ages. *Journal of Genetic Psychology, 131,* 243-253.

Krantz, M. (1982). Sociometric awareness, social participation, and perceived popularity in preschool children. *Child Development, 53,* 376-379.

Krebs, D. L. (1975). Empathy and altruism. *Journal of Personality and Social Psychology, 32,* 1134-1146.

Krebs, D. L. (1980, August). *On the relationship between role-taking and altruism.* American Psychological Association, Montreal.

Krebs, D., & Gilmore, J. (1982). The relationship among the first stages of cognitive development, role-taking abilities, and moral development. *Child Development, 53,* 877-886.

Krebs, D., & Russell, C. (1981). Role-taking and altruism: When you put yourself in the shoes of another, will they take you to their owner's aid? In J. P. Rushton & R. M. Sorrentino (Eds.), *Altruism and helping behavior.* Hillsdale, NJ: Lawrence Erlbaum.

Kurdek, L.A. (1978a). Perspective taking as the cognitive basis of children's moral development: A review of the literature. *Merrill-Palmer Quarterly, 24,* 3-28.

Kurdek, L. A. (1978b). Relationship between cognitive perspective taking and teachers' ratings of children's classroom behavior in grades one through four. *Journal of Genetic Psychology, 132,* 21-27.

Kurdek, L. A. (1980). Developmental relations among children's perspective taking, moral judgment, and parent-related behaviors. *Merrill-Palmer Quarterly, 26,* 103-121.

Lieman, B. (1978, August). *Affective empathy and subsequent altruism in kindergarteners and first graders.* Paper presented at the meeting of the American Psychological Association, Toronto.

Levine, L. E., & Hoffman, M. L. (1975). Empathy and cooperation in four-year-olds. *Developmental Psychology, 11,* 533-534.

Maccoby, E. E. (1980). *Social development: Psychological growth and the parent-child relationship.* New York: Harcourt Brace Jovanovich.

Marcus, R. F. (1980). Empathy and popularity of preschool children. *Child Study Journal, 10,* 133-145.

Marcus, R. F., Telleen, S., & Roke, E. J. (1979). Relation between cooperation and empathy in young children. *Developmental Psychology, 15,* 346-347.

Marsh, D. T., Serafica, F. C., & Barenboim, C. (1981). Interrelationships among perspective taking, interpersonal problem solving, and interpersonal functioning. *Journal of Genetic Psychology, 138,* 37-48.

McGuire, K. D., & Weisz, J. R. (1982). Social cognition and behavior correlates of preadolescent chumship. *Child Development, 53,* 1478-1484.

Mead, G. H. (1934). *Mind, self, and society.* Chicago: University of Chicago Press.

Mehrabian, A., & Epstein, N. (1972). A measure of emotional empathy. *Journal of Personality, 40,* 525-543.

Murphy, L. B. (1937). *Social behavior and child personality.* New York: Columbia University Press.

Mussen, P., & Eisenberg-Berg, N. (1977). *Roots of caring, sharing, and helping: The development of prosocial behavior in children.* San Francisco: W. H. Freeman.

Perry, D. G., Bussey, K., & Freiberg, K. (1981). Impact of adults' appeals for sharing on the development of altruistic dispositions in children. *Journal of Experimental Child Psychology, 32,* 127-138.

Piaget, J. (1926). *The language and thought of the child.* New York: Harcourt, Brace.

Piaget, J. (1965). *The moral judgment of the child.* New York: Free Press. (Original work published 1932.)

Piaget, J. (1967). *Six psychological studies.* New York: Random House.

Piaget, J., & Inhelder, B. (1956). *The child's conception of space.* London: Routledge & Kegan Paul.

Piaget, J., & Inhelder, B. (1969). *The psychology of the child.* New York: Basic Books.

Radke-Yarrow, M., & Zahn-Waxler, C. (in press). Roots, motives, and patterns in children's prosocial behavior. In J. Reykowski, J. Karylowski, D. Bar-Tal, & E. Staub (Eds.), *Origins and maintenance of prosocial behaviors.* New York: Plenum.

Rothenberg, B. B. (1970). Children's social sensitivity and the relationship to interpersonal competence, intrapersonal comfort, and intellectual level. *Developmental Psychology, 2,* 335-350.

Rubin, K. H. (1972). Relationship between egocentric communication and popularity among peers. *Developmental Psychology, 7,* 364.

Rubin, K. H. (1973). Egocentrism in children: A unitary construct? *Child Development, 44,* 102-110.

Rubin, K. H. (1978). Role-taking in childhood: Some methodological considerations. *Child Development, 49,* 428-433.

Rubin, K. H. (1982). Nonsocial play in preschoolers: Necessarily evil? *Child Development, 53,* 651-657.

Rubin, K. H., & Schneider, F. W. (1973). The relationship between moral judgment, egocentrism, and altruistic behavior. *Child Development, 44,* 661-666.

Sawin, D. (1979, March). *Assessing empathy in children: A search for an elusive construct.* Paper presented at the meeting of the Society for Research in Child Development, San Francisco.

Selman, R. L. (1971). Taking another's perspective: Role-taking development in early childhood. *Child Development, 42,* 1721-1734.

Selman, R. L. (1976). Social-cognitive understanding: A guide to educational and clinical practice. In T. Lickona (Ed.), *Moral development and behavior: Theory, research, and social issues.* New York: Holt, Rinehart & Winston.

Selman, R. L., & Byrne, D. F. (1974). A structural-development analysis of levels of role-taking in middle childhood. *Child Development, 45,* 803-806.

Shantz, C. U. (1975). The development of social cognition. In E. M. Hetherington (Ed.), *Review of child development research: Vol. 5.* Chicago: University of Chicago Press.

Spivack, G., Platt, J. J., & Shure, M. B. (1976). *The problem-solving approach to adjustment.* San Francisco: Jossey-Bass.

Spivack, G., & Shure, M. (1974) *Social adjustment of young children.* San Francisco: Jossey-Bass.

Staub, E. (1971). The use of role-playing and induction in children's learning of helping and sharing behavior. *Child Development, 42,* 805-816.

Staub, E. (1978). *Positive social behavior and morality: Social and personal influences: Vol. 1.* New York: Academic Press.

Staub, E. (1979). *Positive social behavior and morality: Socialization and development: Vol. 2.* New York: Academic Press.

Underwood, B., & Moore, B. (1982). Perspective-taking and altruism. *Psychological Bulletin, 91,* 143-173.

Weinstein, E. A. (1969). The development of interpersonal competence. In D. A. Goslin (Ed.), *Handbook of socialization theory and research.* Chicago: Rand McNally.

White, J. B., & Feldman, R. S. (1981, April). *Children's decoding ability and role-taking skills.* Paper presented at the meeting of the Society for Research in Child Development, Boston.

Yarczower, M., & Daruns, L. (1982). Social inhibition of spontaneous facial expressions in children. *Journal of Personality and Social Psychology, 43,* 831-837.

Yarrow, L. J. (1979). Emotional development. *American Psychologist, 34,* 951-957.

Zahn-Waxler, C., & Radke-Yarrow, M. (1979, March). *A developmental analysis of children's responses to emotions in others.* Paper presented at the meeting of the Society for Research in Child Development, San Francisco.

Zahn-Waxler, C., & Radke-Yarrow, M. (1982). The development of altruism: Alternative research strategies. In N. Eisenberg-Berg (Ed.), *The development of prosocial behavior.* New York: Academic Press.

3

Comforting
Communication

Brant R. Burleson

Life holds hurts and disappointments for all of us: the roman-
tic relationship that didn't quite work out, the important promotion we didn't get, the "friend" who once accepted us but now
seems to reject us. Children's lives are also touched by disappointments and hurts. The experience of being excluded from a party
or game, of being the brunt of a cruel remark or name, or of failing an important assignment in school are common in the world
of the child. Although these mundane events may only rarely leave
permanent scars on those they affect, they do create at least temporary feelings of rejection, anger, and sadness. And unless there
is a caring other who attempts to help the distressed child (or adult)
overcome his or her despondent feelings, these painful events may
have more than a merely temporary impact. Fortunately, not only are children capable of recognizing distressed feelings in others,
but they also seem quite willing to provide sympathy and comfort to those experiencing emotional hurt. This chapter is concerned with how children and adults try to help emotionally distressed others overcome their despondent feelings. The concern
here is particularly with the verbal skills used in efforts to alleviate

the emotional hurts experienced by others, the development of these verbal skills, and some of the variables related to the development and use of these verbal skills.

THE NATURE OF COMFORTING

Comforting communication can be defined as the type of communicative behavior having the intended function of alleviating, moderating, or salving the distressed emotional states of others. As such, comforting communication is directed at managing or modifying the psychological states of others. While physical or material conditions may be responsible for producing states of emotional distress, the conception of comforting offered here assumes that the focus of communicative efforts is the management of distressed *feelings*, rather than the remedying of physical injuries or deleterious material conditions. Moreover, although the comforting of a distressed other can certainly be effected through nonverbal means (for example, a compassionate look or a warm hug), the primary focus of this chapter is on the verbal skills people use in seeking to provide comfort. It should also be noted that our concern is with the skills people use to address the moderate feelings of disappointment or sadness arising from everyday events (for example, a child feeling sad about not being invited to a party, or a college student feeling upset about not receiving an academic award). Strategies used to cope with extreme feelings of depression or grief arising from extraordinary events (such as the loss of a spouse) are not dealt with in this chapter.

Viewed as the management of everyday disappointments and hurts, comforting communication represents a theoretically and pragmatically interesting form of behavior. Theoretically, comforting may be viewed both as a type of functional communication skill and as a type of prosocial behavior. Allen and Brown (1976) regard functional communicative competencies as those abilities directed at the accomplishment of certain social objectives (informing others, persuading others, controlling others, and so on). Prosocial or altruistic behavior refers to actions carried out with the primary intention of benefiting another without any expectation of receiving rewards from external sources (Macaulay & Berkowitz, 1970; Mussen & Eisenberg-Berg, 1977). Forms of prosocial behavior include sharing, donating, rescuing, cooperating, helping, and sympathizing.

Functional communication skills and prosocial actions have typically been viewed as quite distinct forms of behavior. However, comforting can be seen as the intersection of functional communication and prosocial behavior. Comforting has a clear social objective (improving the affect state of another) and is usually enacted with the primary intention of benefiting another. Thus, it may be usefully regarded as a type of prosocial communication. Viewing comforting in this manner enables researchers to draw upon the extensive (but unintegrated) literature that has developed with respect to functional communication skills and prosocial behavior. A word of caution is in order here: Several studies (for example, Payne, 1980) have found assessments of different prosocial behaviors (that is, sharing and helping) to be only weakly correlated, and similar results have been reported for assessments of different functional communication skills (for example, Burleson & Delia, 1983). Consequently, although research on other types of prosocial behavior and functional communication skills may be suggestive with respect to comforting, it cannot be taken for granted that comforting skills are related to the same variables as other prosocial behaviors and communication skills.

Comforting behavior is important for several pragmatic reasons. Among groups of children, certain types of functional communication skills, as well as certain types of prosocial activity, have been found to be positively related to social acceptance and popularity (Dlugokinski & Firestone, 1973; Rubin, 1972). In particular, research by Spivack, Platt, and Shure (1976) suggests that social skills such as comforting are related to social adjustment and acceptance. These are especially significant findings since longitudinal studies (Cowen, Pederson, Babigian, Izzo, & Trost, 1973; Roff, Sells, & Golden, 1972) have found that unpopular and unaccepted children are much more susceptible later in life to a host of psychological and social adjustment problems. These longitudinal findings have led to efforts to train unpopular children in social interaction and communication skills (Ladd, 1981; Oden & Asher, 1977). These training studies have found that long-term gains in social acceptance can be effected through experimental interventions. All of this suggests that comforting skills may be an important determinant of social acceptance. Once the nature of effective comforting is better understood, it may be possible to develop training programs designed to enhance comforting abilities.

Despite its theoretical and pragmatic significance, comforting has received relatively little attention compared with other forms of functional communication and prosocial behavior. Indeed, it is only in the last few years that researchers have begun to examine comforting behavior. The remainder of this chapter reviews this small but growing body of research. Methods that have been employed in the study of comforting are discussed initially. Subsequent sections consider individual differences in comforting skill, the effect of situational variables on comforting acts, and some of the factors that lead to the development of comforting skills. The concluding section proposes a general model of the comforting process and sketches some directions for future research.

METHODS FOR STUDYING
COMFORTING BEHAVIOR

The study of comforting communication raises methodological problems not encountered with other functional communication skills or other types of prosocial behavior. For example, laboratory studies of referential communication have frequently operationalized referential skills in terms of the number of accurate choices elicited by a speaker's messages (for example, Krauss & Glucksberg, 1969). However, the use of such "effects" or behavioral outcome measures in the study of comforting is precluded by both ethical and practical considerations: Generating a sample of distressed persons for speakers to comfort is obviously problematic on ethical grounds, and there does not appear to be a practical way of assessing objectively the actual effectiveness of comforting strategies directed toward distressed others. Many studies of prosocial behavior have operationalized the dependent variable in terms of the number of candies, tokens, or pennies donated in response to some experimental task (for example, Rubin & Schneider, 1973); however, comforting activity is not particularly amenable to this sort of pure quantitative analysis. As a result of its special character, nearly all studies of comforting have employed one of three data collection procedures (naturalistic observation, experimentally structured situations, and structured interviews) and one of three methods for coding observed comforting acts (frequency counts, behavioral ratings, and content analysis).

Methods of Data Collection

Naturalistic observation. The most straightforward and ecologically valid way of studying comforting behavior relies on observing spontaneous "real-world" acts of comforting as they occur in natural settings. Once a definition of a comforting act has been established, researchers simply observe subjects in their natural state and record comforting acts as they occur. In spite of its conceptual simplicity and high validity, naturalistic observation is a demanding and very expensive data collection procedure. In addition. logistical and practical concerns make the observation of any "natural" behavior by adults problematic. More particularly, social pressures to maintain a postive public "face" (Goffman, 1967) make it unlikely that scientific observers would be privy to many real-world acts of comforting by older children or adults. These reasons may explain why only one naturalistic observation study of adult comforting behavior has been reported (Applegate, 1980a); moreover, this study focused on the comforting strategies directed by adults at young children.

There are many more opportunities to observe children than adults, and young children are not as conscious of social pressures to maintain a positive face. As a result, there have been several naturalistic observation studies of comforting by young children. However, many of these (for example, Eisenberg-Berg & Hand, 1979; Iannotti, 1981; Yarrow & Waxler, 1976) have reported that the frequency of naturally occurring comforting acts was so low as to preclude meaningful analyses. For example, Iannotti (1981) reports that during a five-month period in which 52 young children were observed for thirty minutes each, 421 naturally occurring prosocial events were recorded, but only one of these was classified as an instance of comforting behavior (also see Strayer, 1981). Whether the observed low frequency of comforting acts in natural settings is due to coding problems or to the inherent rarity of these acts is not entirely clear; however, the available evidence suggests that naturalistic observation is of limited utility in the study of comforting.

Structured situations. In an effort to overcome the limitations associated with naturalistic observation, some researchers have employed experimentally structured situations in which a confederate feigns distress in the hope of eliciting comforting behavior from subjects. For example, Yarrow and Waxler (1976) observed children's reactions to the distress signals emitted by an adult confederate in two structured situations. In one situation, the confederate began sobbing and sniffling while reading an ostensibly sad story. In the other situation, the confederate feigned pinching a finger in a drawer, grimaced, and then proceeded to run cold

water over the "injured" finger. The use of confederates has also
been employed in the study of adults' comforting skills. For example, Feinberg (1977) constructed a situation in which female
college students interacted with an "emotionally distressed"
female confederate. The confederate and the subject were placed
in a room and instructed to complete several personality tests. One
of the tests supposedly reminded the confederate that she had
recently been dropped by her long-term boyfriend. The confederate then began stating that she was upset, hurt, and felt
miserable. Subjects' reactions to the "distressed" confederate were
observed and recorded by concealed experimenters.

Using experimentally structured situations to elicit comforting
behaviors offers several advantages. As suggested by the
naturalistic observation studies discussed above, publicly observable displays of spontaneous comforting activity are rather infrequent. Structured situations thus enable researchers to elicit
a form of behavior that is rarely available to the scientific eye. In
addition, the use of structured situations permits researchers to
control factors that are believed to influence the character of comforting activity.

There are, however, several potential problems associated with
the use of experimentally structured situations. As Rushton (1980)
indicates, an experimental situation can be viewed as a single-item
test. Viewed as such, questions may be raised about the reliability and validity of such tests. Ideally, researchers should employ
multiple items (that is, multiple experimentally structured situations) in eliciting comforting behavior in order to maximize the
reliability of their assessments. Unfortunately, several practical
considerations militate against the use of multiple experimental
situations. The validity of results obtained through the use of experimental situations is obviously dependent on subjects believing that these situations are "real," and it seems unlikely that
relatively sophisticated subjects (older children, adolescents, and
adults) would believe that a context in which they sequentially encountered several distressed others had much reality to it. All of
these considerations suggest that, in spite of the desirability of
employing multiple experimental situations, only one situation (or
at the most, a very few) can be practically used in a single study.
This means that researchers relying on this data collection procedure must take great care to ensure that the employed experimental situation is natural, is likely to have a high "reality factor," and is representative of the class of situations about which
they are interested in making generalizations.

Interviews. A third procedure for collecting examples of comforting behavior involves the use of "clinical" or semistructured interviews (for detailed discussions of this data collection procedure, see Damon, 1977; Selman, 1980). Researchers using this procedure typically present subjects with several hypothetical but realistic situations involving a distressed other and then ask subjects what they would say to make the other feel better. For example, children have been asked to state what they would say to make a friend feel better about not being invited to another child's party (Burleson, 1982a), and college students have been asked to state what they would say to make a roommate feel better about not winning an academic scholarship (Applegate, 1978). In addition, subjects frequently have been asked to supply rationales or justifications for their choices of comforting strategy.

The interview offers several advantages as a data collection technique. First, it is more economical to collect a corpus of examples of comforting behavior through interviews than through either naturalistic observation or experimentally structured situations; it is even possible to mass-administer written forms of the interview schedule to older populations. Second, reliance on standard hypothetical situations offers a measure of experimental control since all subjects respond to some situations, and researchers can tailor these situations to meet particular theoretical concerns. Third, because subjects are aware that they are responding to hypothetical rather than real situations, multiple stimulus situations can be employed without fear of straining subject credulity. This has obvious advantages in terms of enhancing test reliability. Research findings indicate that tests employing multiple hypothetical situations provide very reliable assessments of comforting skills. Internal consistencies of .80 and higher have been reported for multiple-situation tests of comforting skills in both child and adult populations (Applegate, Burke, Burleson, Delia, & Kline, in press; Burleson, 1982a, in press).

Although interviews appear to produce highly reliable results, questions may be raised about the validity of these results. Specifically, it may be asked whether the comforting behavior elicited in response to hypothetical situations is comparable to the comforting behavior exhibited by persons in real-world situations. Applegate (1980a) reports findings directly pertinent to this issue. In his study, a group of daycare teachers responded to several hypothetical situations involving emotionally distressed children. Applegate also observed the teachers' interactions with children at the daycare center over a three-month period, specifically noting natural comforting efforts. Results indicated that teachers employing highly sensitive comforting strategies during the interview tended to use highly sensitive strategies in real-world settings, although they occasionally employed less sensitive strategies as well. On the other hand, teachers employing less sensitive comforting strategies in the interview context consistently employed less sensitive strategies in the natural contexts. This pattern of results generally supports the validity of the interview procedure as a method of assessing comforting skills. In addition, the fact that the teachers who employed highly sensitive strategies in the interview context tended to use both highly sensitive and less sensitive strategies in natural contexts suggests that responses obtained in interviews can be viewed as defining an individual's level of comforting *competence* (that is, the maximum level of ability). A variety of factors in natural situations may result in individuals using comforting strategies less sensitive than they are capable of using (for a more detailed discussion of this issue, see Applegate, 1980a). Further evidence supporting the general validity of the interview procedure is provided by comparative studies of performance in interview and natural contexts with regard to persuasive communication (Applegate, 1982) and interpersonal understanding (Selman, 1980).

Although interviews can provide both reliable and valid assessments of comforting skills, several qualifications regarding this data collection procedure should be noted. First, the hypothetical situations used to elicit comforting strategies must be salient for the population under study. This means that different situations will normally be required for different populations (for example, children versus adults). Second, assuming that multiple situations are used to elicit comforting behaviors, care should be taken to ensure that they meet accepted psychometric standards (for example, similarity of means and standard deviations). Third, it is important to note that interviews *cannot* be used to investigate the comforting skills of all populations. In pilot work, Samter (1982) recently found that preschool-aged children were not able to retain crucial elements of presented hypothetical situations or to form a coherent response to these situations.

Methods of Data Coding

Three basic procedures have been employed to code or evaluate obtained comforting responses. These include frequency counts, behavioral ratings, and content analysis.

Frequency counts. Perhaps the simplest method of evaluating comforting behavior involves counting the number of examples of this behavior observed in natural or experimental situations. For example, naturalistic studies of comforting behavior have often operationalized a comforting predisposition in terms of the number of comforting acts observed during a fixed time period (Bar-Tal, Raviv, & Goldberg, 1982). In an effort to improve the validity of such frequency counts or behavior rates, some studies (for example, Strayer, 1981) have controlled for the number of situations in which a subject was asked to comfort distressed others. The assumption underlying such frequency counts appears to be that individuals displaying a greater number of comforting acts in a given time period are generally more predisposed to comfort the distressed.

Although frequency counts have most often been used in naturalistic observation studies of comforting, they have also been employed to evaluate comforting behavior elicited in response to experimental and hypothetical situations. In her study of comforting behavior exhibited by adult females, Feinberg (1977) counted the number of times a subject smiled at the "distressed" confederate. Burleson (1982a) employed four hypothetical situations to assess developmental changes in children's and adolescents' comforting skills. This researcher counted the number of different strategies and the variety of strategies employed by subjects in responding to each situation. It was argued that individuals employing a greater number and diversity of comforting acts had a broader repertoire of strategies to draw upon, and thus could be expected to be more flexible in responding to the distress of others.

Three issues are of central import in the use of frequency counts. First, the definition of what constitutes a comforting act is crucial. Care must be taken to ensure that operationalizations of comforting acts are at least face-valid, and that there is high reliability in the coding of such acts. Second, researchers should not assume that the theoretical significance of various frequency counts is transparent. What it means for an individual to produce more of a given behavior, or to produce that behavior at a higher rate, may not be entirely clear. Thus, attention should be given to the assumptions underlying the use of specific frequency counts. Finally, the frequency with which a particular behavior appears, especially in natural settings, may be influenced by a host of factors. Researchers must be aware of these factors and take them into account in interpreting their results.

Behavioral ratings. A second method employed to evaluate comforting behavior utilizes predetermined dimensions along which the observed behavior is rated. For example, to evaluate children's responses to the feigned distress of adult confederates, Yarrow and Waxler (1976) rated the observed behavior using a six-point scale for the degree of comforting and empathic involvement. Feinberg (1977) employed a series of rating scales to assess the quality of the comforting behavior directed by adults at a confederate. Subjects were rated for their responsiveness, attentiveness, perceived concern, perceived help, and perceived sympathy. Ratings were made by both the confederate and concealed observers.

Behavioral ratings have a number of significant advantages. Rating scales are both economical and fairly easy to use. High degrees of intercoder reliability can be obtained quickly with many rating scales. Rating scales also tend to be very flexible and are readily applicable to data collected through naturalistic observation, experimentally structured situations, or interviews. General behavioral ratings may be especially useful in early phases of research when the specific features of the behavior under study are not well understood, or when some holistic evaluation of exhibited behavior is desired.

While behavioral ratings have appropriate applications, there are also a number of significant limitations associated with this method of evaluating comforting acts. The specific scales selected to rate examples of comforting behavior have often had a certain ad hoc quality to them. The choice of a rating scale frequently appears to have been based on intuition rather than any specific theoretical criteria. Although researchers may be forced to rely on their intuitions initially, it is vital that the basis of these intuitive judgments be articulated as precisely as possible so that more rigorous coding methods can be developed. Moreover, although ratings of "helpfulness," "concern," and "sympathy" are useful for certain purposes, such global ratings do little to aid our understanding of the precise constituents of a "helpful" or "sympathetic" communicative strategy. Knowledge of such constituents is necessary both for theoretical development and pedagogical purposes.

Content analysis. A third approach to the evaluation of comforting acts involves the coding of comfort-intended strategies based on the presence (or absence) of specific message features.

Specific content characteristics of messages are used to define a set of categories, and observed messages are then coded using these catetories. Although the categories composing a content analytic system do not have to be arrayed in any particular way (see Feinberg, 1977), most studies of comforting behavior have employed coding systems in which an ordinal or hierarchical relationship among the categories has been assumed (for a general discussion of hierarchical content analysis, see Clark & Delia, 1979). For example, Applegate (1980b; also see Burleson, 1982a) developed a nine-level hierarchical coding system that scores comforting strategies for the extent to which they deny, implicitly recognize, or explicitly articulate and elaborate the feelings and perspective of a distressed other (see Table 3.1).

Such hierarchical coding systems have a clear advantage: Strategies not only can be differentiated on the basis of specific message features, but it is also possible to assert that one strategy is "better" (more sophisticated, adaptive, sensitive, and so on) than another on the basis of its hierarchical placement. However, it must be clearly understood that such qualitative judgments are made strictly within the confines of the principle underlying the ordering of the categories. For example, one could not infer (in the absence of empirical support) that comforting strategies defined as more sensitive in Applegate's coding system are also generally more effective than less sensitive strategies.

Hierarchically ordered coding systems can be viewed as Guttman scales, in that the use of a particular strategy level implies the capacity to use all strategy levels lower on the hierarchy. For example, a person capable of using a high-level comforting strategy explicitly elaborating another's feelings (a level eight strategy; see Table 3.1) might employ a strategy diverting the other's attention from the distressful situation (a level four strategy) if that person thought the other's distressed state was relatively transient and that dwelling on that state would only exacerbate the situation. Although reversion to the use of lower-level strategies is a possibility, the converse should not be true. Moreover, studies in which independent assessments of message strategies and message rationales have been obtained indicate that relatively little reversion to the use of lower-level strategies takes place, at least in the structured interview context (Applegate, 1980b). Finally, although a subject's response to a stimulus situation may contain multiple comforting strategies codable within different hierarchical levels, most research has focused on the

TABLE 3.1   Hierarchical Coding System for Sensitivity of Comforting Strategies

"What would you say to make a friend feel better about not receiving an invitation to another child's party?"

I. Denial of Individual Perspectivity

The speaker condemns or ignores the specific feelings that exist in the situation for the person addressed. This denial may be either explicit or implicit.

   1. Speaker condemns the feelings of the other:

      "I'd tell her she had no reason to feel that way about not getting invited, and if she felt that way, she was no friend of mine."

   2. Speaker challenges the legitimacy of the other's feelings:

      "There's nothing to be upset about—it's just an old party."

   3. Speaker ignores the other's feelings:

      "I'd tell her there've been other parties, and she should be happy about going to them."

II. Implicit Recognition of Individual Perspectivity

The speaker provides some implicit acceptance of and/or positive response to the feelings of the others, but does not explicitly mention, elaborate, or legitimize those feelings.

   4. Speaker attempts to divert the other's attention from the distressful situation and the feelings arising from that situation:

      "When it's my party, I'll invite you."

   5. Speaker acknowledges the other's feelings, but does not attempt to help the other understand why those feelings are being experienced or how to cope with them:

      "I'm sorry you didn't get invited to the party. I'm sorry you feel bad."

   6. Speaker provides a non-feeling-centered explanation of the situation intended to reduce the other's distressed emotional state:

      "Maybe your invitation to the party got lost in the mail. Or maybe there just wasn't enough room to invite everybody in the whole class."

III. Explicit Recognition and Elaboration of Individual Perspectivity

The speaker explicitly acknowledges, elaborates, and legitimizes the feelings of the other. These strategies may include attempts to provide a general understanding of the situation. Coping strategies may be suggested in conjunction with an explication of the other's feelings.

   7. Speaker explicitly recognizes and acknowledges the other's feelings, but does not discuss the source of the feelings or the nature of the situation:

      "I know you feel bad about not going to the party, but you're my friend—lots of people like you. When my party comes up, I'll invite you."

(continued)

TABLE 3.1 Continued

8. Speaker provides an elaborated acknowledgment and explanation of the other's feelings:

   "Gee, I'm really sorry about the party. I didn't mean to make you feel bad by mentioning it, but I know I did. It's not fun being left out. Maybe it's a mistake. I'll talk to Sharon. OK?"

9. Speaker helps the other to gain a perspective on his or her feelings and attempts to help the other see these feelings in relation to a broader context or the feelings of others:

   "Well, I'd tell her that I really understand how she feels, that I haven't been invited to a special party sometimes and I know it hurts—you can feel rejected. But I'd say maybe Jean really wanted to have you, but her parents wouldn't let her invite everybody. And that I've had parties where I couldn't invite everybody I wanted, and she probably has too. So it doesn't mean that Jean doesn't like her or anything, just maybe her mom was letting her have like only a few people."

SOURCE: Adapted from Applegate (1980b) and Burleson (1982a).

NOTE: Roman numerals correspond to major levels of the coding system. Arabic numerals correspond to the sublevels at which specific message strategies were scored.

highest-level strategy used in responding to the situation. This is because most studies employing hierarchical codings have endeavored to assess subjects' competence to engage in sensitive comforting, and, as Loevinger and Wessler (1970) suggest, the highest-level scoring algorithm is appropriate when the characteristic under study is viewed primarily as capacity.

Different hierarchical coding systems based on various abstractive principles can be constructed and applied to the same strategies, thereby providing a multifunctional analysis of comforting behavior (O'Keefe & Delia, 1982). For example, Applegate (1980b) not only coded comforting strategies for the extent to which the feelings of the distressed other were elaborated, but he also employed a different hierarchical system to code these strategies for the extent to which they implicitly sought to preserve and enhance a positive interpersonal relationship with the other. More recently, Applegate et al. (in press) utilized Hoffman's (1977a) distinction between "power-assertive" and "inductive" appeals to develop a six-level hierarchical system coding parents' comforting strategies for the extent to which a distressed child was encouraged to reflect on his or her feelings and the circumstances producing these feelings.

The procedures employed to develop hierarchical systems for coding message strategies can also be applied to the construction of hierarchical systems for assessing message rationales, the justifications people supply in explaining their message choices. Applegate (1980b) and Burleson (1980) both developed similar hierarchies for assessing the extent to which rationales for comforting strategies reflect concern with specific features of the distressed other and the distressful situation.

Several procedures are available for validating a hierarchical coding system. The face-validity of the coding system is determined by whether the categories and the hierarchical arrangement of these categories instantiate the abstractive principle. Construct validation strategies depend on the specific character of the abstractive principle underlying the hierarchy. For example, many of the hierarchical systems used to assess aspects of comforting behavior are derived from a "constructivist" view of cognitive-developmental theory (see Applegate & Delia, 1980; Delia, O'Keefe, & O'Keefe, 1982). Thus it was expected that older children would employ higher-level strategies than their younger counterparts, a prediction borne out by several empirical investigations (see below). Other evidence supporting the construct validity of hierarchical coding systems for various types of comforting behavior is reviewed in subsequent sections of this chapter.

INDIVIDUAL DIFFERENCES IN COMFORTING BEHAVIOR

This section reviews findings on differences in comforting behavior as a function of such person variables as age, sex, social class, general cognitive abilities, social perception skills, and certain personality traits. Although some researchers (Gergen, Gergen, & Meter, 1972; Rosenhan, Moore, & Underwood, 1976) have maintained that variations in prosocial activities are primarily due to situational factors, the research reviewed below indicates that stable personal qualities substantially influence one's comforting skills.

Age

Several related literatures lead to the expectation that comforting skills will increase over childhood and adolescence. The frequency and quality of many other prosocial behaviors (for example, helping, sharing, and cooperating) have been found to increase with age (see Rushton, 1976; Staub, 1979). Similarly, considerable evidence indicates that many functional communication skills exercised by children (referential or persuasive skills) undergo marked advances during childhood and adolescence (see Asher, 1979; Delia & O'Keefe, 1979; and Glucksberg, Krauss, & Higgins, 1975). In a related vein, children's capacities to recognize and manage interpersonal problems have been found to increase with age (Spivack et al., 1976).

Spontaneous acts of comforting behavior have been observed in children as young as one. Zahn-Waxler, Iannotti, and Chapman (1982) studied infants' (nine to thirty months) reactions both to the natural distresses of others and to simulated distresses enacted by experimenters and caregivers. These researchers found that even the youngest infants in their study (nine to fourteen months) directly intervened on behalf of others in a measurable percentage of situations, with the frequency and types of comforting behavior reflecting developmental trends.

Several naturalistic observation studies of preschoolers (Bar-Tal et al., 1982; Iannotti, 1981; Murphy, 1937; Strayer, 1981; Yarrow & Waxler, 1976) suggest that the tendency to provide comfort to distressed others increases between three and six years of age. For example, Bar-Tal et al. (1982) observed 156 children ranging in age from 18 to 76 months during free play periods at a daycare center. The number of children observed performing at least one "real" comforting act (that is, an act performed in a real rather than a play situation) was found to increase significantly with age. However, even among the oldest children examined (67-76 months), comforting acts were relatively rare: Only 29 percent of the children in the oldest group were observed performing comforting acts. These findings appear to be consistent with several other naturalistic studies of preschoolers (for example, Murphy, 1937). These studies indicate that although preschoolers engage in comforting acts relatively infrequently, both the number of children engaging in comforting acts and the occurrence of these acts increase with age.

Yarrow and Waxler (1976) observed the reactions of 108 children (36 to 90 months in age) to the distress signals emitted by an adult confederate in two experimental situations. As noted above, one situation involved the confederate crying over an ostensibly sad story, and the other situation had the confederate feign pinching a finger. The quality of children's comforting responses to the confederate, assessed on a six-point rating scale, were found to be negatively correlated with age at a significant level ($r = -.29$). However, Yarrow and Waxler noted that this negative association was solely due to the older children responding less solicitously to the adult's pinched finger. Several factors may be responsible for this somewhat surprising finding. In particular, the authors suggest that the older children may have perceived the adult as being better able to handle the situation of the pinched finger than they were, while the younger children may have spontaneously acted on the basis of helpful intentions without considering if they would be able to offer any useful aid. These findings indicate that the characteristics of the distressed other or the distressful situation may significantly influence comforting efforts.

Three studies have examined developmental changes in the comforting message strategies employed by school-aged children and adolescents. Ritter (1979) asked high school freshmen and seniors to imagine that they had unintentionally hurt the feelings of a peer by publicly boasting about how easily they had passed a driving test that, unknown to them, the peer had failed. Subjects were instructed to suppose they had learned of the peer's hurt feelings and were asked to state what they would say to make the peer feel better. Messages produced in response to this situation were coded within hierarchical schemes similar to those developed by Applegate (1980b). Ritter found that, compared with freshmen, seniors used comforting strategies providing significantly greater acknowledgment of and responsiveness to the feelings of the distressed peer.

Burleson (1982a) replicated and extended Ritter's work, examining several aspects of the comforting strategies employed by first through twelfth graders. Subjects in this study were asked to respond to four hypothetical situations in which peers were depicted as experiencing varied forms of emotional distress (for example, sadness, fear, or anger). Responses to each situation were coded for the number of strategies employed, the variety of strategies employed, and the extent to which strategies acknowledged, elaborated, and legitimized the feelings and perspective of the distressed peer. Strong linear relationships were found between age and the number, variety, and sensitivity of strategies employed. Burleson interpreted these findings as indicating that children's repertoires of comforting strategies undergo both quantitative and qualitative changes during childhood and adolescence.

In an earlier study, Burleson (1980) examined the rationales employed by children and adolescents in explaining their choices of specific comforting strategies. Children from the second, fourth, sixth, and eighth grades were presented with hypothetical situations involving an emotionally distressed peer, were asked what they would say to make the peer feel better, and then were asked to explain *why* what they had said would make the peer feel better. Rationales for the comforting messages were coded within a six-level hierarchy for the extent to which message choices were based on an integrated consideration of the other's personality characteristics and the salient features of the situation. A strong age effect was detected: With increasing age, children employed rationales for their comforting efforts that reflected progressively greater awareness of the specific characteristics of the distressed other and how these characteristics should be dealt with in the comforting process.

Although there are noticeable gaps in the literature, existing studies complement one another in suggesting that comforting skills develop throughout childhood and adolescence. Rudimentary acts of comforting appear either during or shortly after the first year of life. By the age of two, many children are engaging in differentiated acts of comforting, and the first verbal expressions of sympathy have appeared. Throughout the preschool years, children become more likely to perform spontaneous acts of comforting, although the frequency of such acts remains relatively rare. By the time children enter grade school, they are quite capable of recognizing states of emotional distress in others and are further capable of formulating more or less appropriate responses to these states. Moreover, the capacity for sensitive comforting continues to develop into late adolescence. Throughout the grade school and high school years, children develop larger repertoires of comforting strategies, employ more sophisticated and sensitive strategies, and give greater attention to personality and situational characteristics in selecting strategies.

Many questions regarding the development of comforting skills remain unanswered. Although comforting skills clearly appear to develop throughout childhood and adolescence, we lack precise understanding as to how and why these skills develop as they do. As a demographic variable, chronological age is at best a summary index of other, more theoretically interesting variables (general cognitive and verbal developments, social cognitive developments, specific social experiences, norm internalization, modeling experiences, and so on). How such variables contribute to the growth of comforting skills needs more intensive study. Several other issues also warrant further investigation. For example, do the developmental changes in comforting skills observed by Burleson (1980, 1982a) and Ritter (1979) occur in a stage-like manner? At what points do major changes in the sophistication of comforting skills occur? Do developmental changes in comforting skills follow a continuous or discontinuous pattern? Answering these questions will require that existing cross-sectional studies be complemented by longitudinal research.

Sex Differences

Few significant sex differences have been reported with respect to comforting skills. However, when such differences have been detected, they have consistently favored females. For example, Burleson (1982a) reports that females used a greater number, a greater variety, and qualitatively more sensitive comforting strategies than did males (also see Borden, 1979; Burleson, 1980). Although these sex differences are statistically reliable, they are not large in magnitude: Burleson (1982a) found that sex accounted for an average of only 3.3 percent of the variance in the comforting skills he studied.

The pattern of sex differences with respect to comforting is consistent with research on other prosocial behaviors and functional communication skills. Small but occasionally significant sex differences favoring females have been reported for general adaptive communication skills (Alvy, 1973) and for several prosocial behaviors (see reviews of Krebs, 1970; and Rushton, 1980). The reason for female superiority with respect to comforting is not entirely clear. Hoffman (1977b) suggests that empathy is a major motivating force for all prosocial behavior and has reviewed several studies indicating that females are more empathic than males. Burleson (1982a) argues that within many Western cultures, females are expected to be "feelings specialists," giving greater attention than males to the psychological and emotional states of others. These cultural expectations may lead females to develop comparatively more sophisticated skills for coping with others' affective states.

Social Class

Socioeconomic status has occasionally been found to be positively associated with prosocial activity (Berkowitz, 1968) and has consistently been found to be related to functional communication skills (Alvy, 1973; Baldwin, McFarlane, & Garvey, 1971; see also Higgins, 1976). Bernstein's (1975) sociolinguistic theory has often been employed in explanations of social class differences in communication. Bernstein argues that distinct modes of speaking are organized by the assumption that either uniqueness or similarity exists in the psychological experiences of interactants. These divergent assumptions about psychological experiences are expressed in two distinct sociolinguistic "codes," which Bernstein labels the "elaborated" (personal) code and the "restricted" (positional) code. He argues that these codes characterize the speech of the middle and lower classes, respectively.

Applegate and Delia (1980) have extended Bernstein's analysis, arguing that social class indexes differences in culturally shared beliefs concerning the nature of persons and social relationships. Life within cultures characterized by a restricted code presumably leads members of the lower classes to view other persons in terms of relatively concrete characteristics (for example, physical qualities and social roles), and to view social relationships as governed by implicit sets of rules. In contrast, life within cultures marked by an elaborated code is assumed to lead middle-class speakers to view others in terms of abstract, psychologically centered dispositions, and to view social relationships as governed by negotiated and emergent understandings.

Applied to comforting communication, Applegate and Delia's analysis suggests that lower-class speakers should employ less sensitive verbal comforting strategies that adapt to distressed others only in global and role-centered ways, while middle-class speakers should employ more sensitive verbal strategies that specifically elaborate on and adapt to the feelings of others. The findings of several studies support this hypothesis. Applegate (1980a) found a strong positive correlation ($r = .75$) between daycare teachers' socioeconomic status and their use of sensitive comforting strategies. In a study of mothers of young children, Jones, Delia, and Clark (1981a) found that socioeconomic status and the sensitivity of both comforting and regulative (disciplinary) strategies were correlated at a moderate level ($r = .52$).

Applegate et al. (in press) also found a significant positive correlation (r = .40) between socioeconomic status and the sensitivity of comforting strategies employed by a different sample of young mothers. However, these researchers argue that the cultural beliefs indexed by social class should impact primarily on the manner in which other persons are cognitively represented (that is, on social cognitive processes), and only mediately—through such representations—on communicative behavior. This hypothesis was assessed through a regression approach to path analysis. When individual differences in social cognitive ability were controlled, socioeconomic status did not make a significant contribution to the prediction of the mothers' communication. In contrast, when controlling for the effect of socioeconomic status, social cognitive ability was found to add substantially to the prediction of their communication. This result provides direct support for the notion that the cultural beliefs captured in social class assessments impact primarily on the ways in which others are psychologically represented, and only mediately on modes of communication. More generally, Applegate et al.'s approach represents a fruitful way of conceptualizing relationships among culture, cognition, and communication.

General Cognitive and Verbal Abilities

Comforting skills appear to be only marginally related to general cognitive and verbal abilities. In a study of young children, Zahn-Waxler et al. (1982) found performance on the Peabody Picture Vocabulary Test to be significantly related to the likelihood of providing comfort to distressed others, but the magnitude of the relationship was not high (r = .33). In another study of children, Burleson and Delia (1983) found a measure of verbal fluency to be associated with the sensitivity of comforting strategies employed by first graders (r = .48); however, these variables were uncorrelated in a sample of third graders. Moreover, these researchers found that the sensitivity of strategies employed by both first and third graders was unrelated to a measure of verbal intelligence. In studies of adults, neither verbal fluency nor verbal intelligence has been found to be associated with comforting skill (Applegate, 1978; Applegate et al., in press). In addition, Borden (1979) found that the number of words used to express a comforting strategy was uncorrelated with the sensitivity of that strategy, as defined by Applegate's (1980b) hierarchical system.

These results suggest that there may be some relation between comforting skills and general cognitive and verbal abilities in young children, but this relation does not appear to persist past middle childhood. This pattern of results is consistent with the view that cognitive and communicative abilities may be fused in early childhood but become progressively differentiated as development proceeds (see O'Keefe & Sypher, 1981).

Social Perception Skills

A relatively large number of studies have examined the relation between comforting behavior and several social perception (social cognitive) skills. It has been argued that social perception skills, especially the capacity to understand another's point of view (that is, role-taking skills), underlie both the tendency to behave prosocially (Krebs & Russell, 1981) and the ability to communicate adaptively (Delia & O'Keefe, 1979; Flavell, 1968).

Presumably, if an individual is to behave prosocially, he or she must be able to recognize or infer that another is in need of aid. Moreover, some forms of effective prosocial intervention appear to depend on understanding how those in need view their situation. A similar line of analysis can be offered for the case of adaptive communication: In order to tailor and edit a message so as to meet listener and situational requirements, a speaker must be able to infer the relevant characteristics of the listener and understand how the listener represents the situation.

As suggested in the first part of this chapter, comforting can be viewed as the intersection of prosocial behavior and adaptive communication skills. Thus, it might be thought that social perception skills are especially relevant to comforting behavior, particularly since the exercise of comforting skills presupposes an awareness of and sensitivity to the emotional states of others. More specifically, the effective comforting of another would appear to require the ability to recognize and understand the cause(s) of the other's distressed state. Several studies have found substantial developmental and individual differences in these specific social perception skills (for example, Burleson, 1982b; Kurdek, 1977; Rubin, 1978).

Despite the intuitive appeal of the posited relationships between social perception skill and both prosocial behavior and adaptive communication, the empirical findings have been mixed. The relationship between social perception and prosocial behavior is far from clear (compare Krebs & Russell, 1981; Kurdek, 1978; Mussen & Eisenberg-Berg, 1977; Underwood & Moore, 1982). Similarly, a highly general relationship between social perception and functional communication skills has not been found (see Delia & O'Keefe, 1979; Shantz, 1981). Perhaps it is understandable, then, that empirical studies of the relationship between social perception skills and comforting behavior have yielded a somewhat inconsistent set of findings.

Social perception and comforting in young children. Several studies have examined the relationship between comforting behavior and different types of role-taking or perspective-taking ability. Theorists have identified three content-related forms of perspective taking (Kurdek, 1978; Shantz, 1975): spatial (inferring what another sees), conceptual (inferring what another knows), and affective (inferring what another feels). Zahn-Waxler, Radke-Yarrow, and Brady-Smith (1977) examined the relationship between all three types of perspective taking and the comforting behavior exhibited by young children in the two experimental situations employed by Yarrow and Waxler (1976). No relationships were found between comforting behavior and any of the perspective-taking measures. Subsequent investigations of the relationship between comforting by young children and varied types of perspective taking have yielded essentially the same result (Eisenberg-Berg & Lennon, 1980; Iannotti, 1981; Strayer, 1980; Zahn-Waxler et al., 1982).

These results suggest that there may not be any relationship between perspective-taking ability and comforting behavior. However, there are other possible explanations. First, recent reviews of the perspective-taking literature (Enright & Lapsley, 1980; Ford, 1979) have suggested that many of the most commonly used perspective-taking tasks may be lacking in reliability and validity. Second, a general level of social perception skill may not be reflected in specific, situated types of behavior. Social perception skills are perhaps best regarded as determinants of the general ability or *competence* to engage in some form of behavior. Most of the studies mentioned above have examined children's comforting in natural situations, contexts where actual behavior may depart from the competence level due to a host of environmental influences.

Third, developed social perception skills such as perspective taking are, at best, a necessary but not sufficient condition for any prosocial behavior or adaptive form of communication. In order to behave prosocially or communicate adaptively, people have to attain a certain level of social perception skill. In addition, they must acquire specific behavioral routines and strategies that enable them to act efficaciously on their perceptions. Applegate and Delia (1980) contend that sophisticated strategic control over behavior develops only after children acquire social perception skills that lead to the formulation of certain behavioral intentions. Moreover, these theorists argue that social perception skills and behavioral resources become effectively integrated only gradually. Empirical support for this viewpoint is reported in a study by Delia and Clark (1977), who found that children passed through a developmental period in which they could clearly identify communicatively relevant differences in listeners, but were unable to formulate a communicative strategy for adapting to these differences. All of the studies discussed above assessing the relationship between perspective taking and comforting employed samples of young children (largely preschoolers). Thus, it is possible that significant relationships were not detected because the children in these studies had not yet fully integrated their cognitive and behavioral capacities.

Social perception and comforting in older populations. Several studies conducted from a constructivist theoretical perspective (Delia et al., 1982) have examined the relationship between comforting behavior and social perception skills. These studies differ from those discussed above in a number of respects. First, the constructivist studies were carried out on older populations (grade school children, adolescents, and adults) in which a fuller integration of cognitive and behavioral capacities could be expected. Second, these studies focused on comforting strategies provided in response to hypothetical situations. As suggested above, responses to hypothetical situations provide perhaps the best estimate of a person's level of behavioral competence. Third, the constructivist studies employed rather different assessments of social perception skills.

The constructivist approach views all social perception processes as occurring through the application of cognitive schemes termed "interpersonal constructs." Following Kelly (1955), interpersonal constructs are viewed as the basic dimensions through which we interpret, evaluate, and anticipate the thoughts and behaviors of others. Systems of interpersonal constructs are known to undergo a series of predictable changes over the course of development. Specifically, with advancing age, construct systems become increasingly differentiated or complex (composed of greater numbers of elements) and comprise a larger proportion of abstract elements (constructs pertaining to traits, motives, and dispositions rather than physical characteristics and social roles; see Barenboim, 1977; Delia, Burleson, & Kline, in press; Scarlett, Press, & Crockett, 1971). Moreover, substantial individual differences in interpersonal construct system complexity and abstractness have been found to be related to a number of specific social perception processes, including impression organization (Delia, Clark, & Switzer, 1974); information integration (Nidorf & Crockett, 1965); social evaluation (O'Keefe & Delia, 1981); and social perspective taking skills (Hale & Delia, 1976). Thus, assessments of construct system complexity and abstractness constitute very general measures of social perception skill. Reliable and valid measures of construct system complexity and abstractness have been developed (see O'Keefe & Sypher, 1981).

Four studies have examined the relationship between indices of construct system development and comforting skill in samples of children. Utilizing an individual-difference research design with first and third graders, Delia, Burleson, and Kline (1979) found that construct system differentiation (but not abstractness) was associated with the sensitivity of comforting strategies produced in response to hypothetical situations. One year later, the same children were retested. Once again, construct differentiation was moderately related to comforting skill. In addition, construct abstractness was found to be significantly associated with comforting skill among the (now) fourth graders. This latter finding was interpreted in light of the fact that few abstract constructs appear prior to middle childhood (see Delia et al., in press; Scarlett et al., 1971).

Ritter (1979) found no relationship between construct differentiation and comforting skill in a sample of high school students. However, this researcher noted that the range of differentiation scores in her sample was quite narrow. Burleson (1981) examined the relationship between comforting strategy sensitivity and both construct differentiation and abstractness in a sample of first through twelfth graders. Even when controlling for the effect of age, level of comforting strategy was moderately associated with both construct differentiation (r = .69) and construct abstractness (r = .41). In addition, comforting skill was found to be highly correlated with performance on a modified version of Hale and Delia's (1976) social perspective taking task (age-partialled r = .70). In a related study, Burleson (1982b) found that the person-centeredness of children's rationales for their comforting strategies was moderately associated with indices of construct differentiation, construct abstractness, and social perspective taking ability.

Seven studies have employed adult samples in examining the relationship between individual differences in construct system development and the quality of comforting strategies used in responding to hypothetical situations. Four of these studies (Applegate, 1978; Borden, 1979; Burleson, 1978, in press) utilized samples of college students. In all four studies, the indices of construct system development were significantly related to the quality of the elicited comforting strategies (average r = .52). The other three studies employed samples of daycare teachers (Applegate, 1980a), primary and secondary teacher trainees (Applegate, 1980b), and mothers of young children (Applegate et al., in press). In each of these studies, construct abstractness was found to be positively associated with the level of comforting skill. However, in one of the studies (Applegate et al., in press), construct differentiation was found to be only marginally associated with comforting strategy sensitivity (r = .24, p = .15).

In summary, studies examining the relationship between individual differences in construct system development and comforting skills have found positive relationships, with correlations usually in the moderate (.40-.60) range. These results are particularly impressive since the measures of construct system development employed have been found to be unrelated to potentially confounding variables such as verbal intelligence and loquacity (Burleson, Applegate, & Neuwirth, 1981; Sypher & Applegate, 1982). The highly consistent pattern of results obtained in these studies indicates that individual differences in social perception skills underlie the capacity to employ sensitive comforting strategies (at least among older childlren and adults). These findings are quite consistent with other research indicating that individual differences in construct system development underlie communicative skills in the referential, regulative, and persuasive domains (Applegate, 1980b; Delia, Kline, & Burleson, 1979; Hale, 1980, 1982; O'Keefe & Delia, 1979).

Personality Traits

A number of studies have explored the relationship between personality traits and prosocial behavior. Staub (1980) suggests that certain personality traits may lead individuals to formulate and pursue prosocial goals. Thus, some personality traits can be viewed as motivating forces for prosocial behavior.

Empathy. Emotional empathy is the personality trait that has been studied most frequently with respect to comforting behavior. Several writers (for example, Batson & Coke, 1981; Feshbach, 1978; Hoffman, 1977b) have suggested that empathizing emotionally with another (that is, feeling what the other feels) is a basic source for the motivation to act prosocially. It should be noted, however, that the trait of emotional empathy has often been confused with the social perception process of affective perspective taking (recognizing and understanding what another feels). Maintaining a distinction between these variables is important since social perception skills are determinants of the competence to comfort others sensitively, while personality traits such as empathy influence the motivation to provide comfort. Moreover, empathy and social perception skills have been found to be unrelated in several studies (for example, Burleson, in press; Eisenberg-Berg & Lennon, 1980). Thus, the present review is concerned only with those studies that view empathy as the experience of feeling what another feels.

Although many studies have examined the relationship between empathy and various types of prosocial behavior (see Underwood & Moore, 1982), only four have focused directly on the relation between empathy and comforting. Eisenberg-Berg and Lennon (1980) examined the relationship between empathy and the rate of comforting acts spontaneously produced by young children in natural settings. No significant relation was found between empathic tendencies and the frequency of comforting acts. These researchers suggest, however, that their measure of empathy (Feshbach and Roe's [1968] Affective Situations Test) may have actually reflected the children's need for approval rather than their empathic tendencies. Zahn-Waxler et al. (1982) also explored the relationship between empathy and children's comforting behavior. Empathic arousal was assessed by rating children's responses to two simulated instances of affective distress. In this case, empathy was found to be negatively associated with comfort given to a distressed adult. However, these writers urge caution in interpreting this result, since empathy was not reliably scored.

Two studies utilizing adult samples have assessed the relation between questionnaire measures of empathy and the quality of comforting strategies elicited in response to hypothetical situations. Borden (1979) found a weak but significant relationship between strategy quality and Hogan's (1969) questionnaire measure of empathy; Burleson (in press) found comforting strategy sensitivity positively associated with Mehrabian and Epstein's (1972) assessment of emotional empathy. These findings suggest that there is some relationship between an empathic disposition and the type of message strategies used to comfort distressed others, at least among adults.

Given the central place occupied by empathy in many theories of prosocial behavior, additional studies of the relation between empathy and comforting need to be undertaken, especially among children. Research with children, however, will require the development of more reliable and valid measures of empathy.

Other personality traits. The impact of additional personality traits on comforting behavior has also been assessed. For example, Borden (1979) assessed the relationship between commitment to interpersonal values and the quality of comforting strategies used in hypothetical situations. A small, but significant, association between these variables was detected.

Feinberg (1977) and Grodman (1979) both created a general personality trait labeled "prosocial orientation" by summing subjects' factor-weighted standardized scores on several personality measures (for example, interpersonal values, Machiavellianism, and ascription of social responsibility). In both studies, female college students interacted with a confederate who appeared distressed about having been dropped by her boyfriend. Subjects with a high prosocial orientation were found to produce more examples of comforting behavior than subjects of a less prosocial orientation. Those high in prosocial orientation looked at the other more, smiled more, generated more helpful verbalizations, and expressed a greater willingness to continue to interact with the confederate. Moreover, those high in prosocial orientation were rated by the confederate and observers as more helpful, attentive, concerned, and sympathetic.

The studies reviewed in this section clearly indicate that individual differences in certain personality characteristics are associated with comforting behavior. Few traits have been studied, however, and none of these have been examined extensively. While much research remains to be done in this area, subsequent investigations will be most beneficial if they are guided by a conceptual model of how personality traits are related to comforting.

SITUATIONAL INFLUENCES ON COMFORTING

Prosocial behavior is known to be affected by a wide variety of situational factors (see Rosenhan et al., 1976). Because of the general paucity of research on comforting, however, the effects of only a few situational factors on comforting behavior have been assessed. Those situational features that have been studied fall roughly into one of three general classes: temporary states of the comforter, characteristics of the distressed other, and characteristics of the distressful situation.

Temporary States of the Comforter

A considerable body of research indicates that temporary states of the actor, especially variations in mood, can substantially affect prosocial behavior (see Barnett, this volume). Nevertheless, only one study has examined the effect of an actor's temporary states on comforting behavior. Applegate (1980a) engaged in three months of naturalistic observation of daycare center teachers, specifically focusing on the comforting and regulative (disciplinary) strategies that these teachers employed with their charges. For each observed incident, Applegate recorded the parties involved, the time of day, activating features of the situations, apparent moods and feelings of the parties involved, what was said, and the outcome of the situation. Several features of the situation appeared to affect the type of comforting strategy employed. For example, teachers appeared to use less sensitive strategies late in the day when they were tired. If teachers appeared angry, frustrated, or anxious about features of the situation, they also tended to employ less sensitive strategies. While these results are not surprising, they are important in documenting the influence of temporary states on displayed comforting skills. Clearly, Applegate's work needs to be followed by more systematic and focused studies.

Characteristics of the Distressed Other

The situational factors receiving the greatest research attention pertain to specific qualities of the distressed other. For example, in Ritter's (1979) study of high school students, the effect of the type of social relationship between the speaker and the distressed other was assessed. Ritter asked her subjects to imagine that they had hurt a peer's feelings by bragging about how easily they had passed a driving test that the peer had failed. In one condition, subjects were asked to suppose that the peer was a close friend, while in the other condition subjects were instructed to imagine that the distressed other was an "out-group" acquaintance. Ritter predicted that more sensitive comforting strategies would be employed with the friend than the acquaintance. This hypothesis was strongly confirmed. These results are consistent with several studies showing that friends behave more prosocially toward one another than they do with nonfriends (see Zahn-Waxler et al., 1982).

The distressed other's need for help has also been shown to influence comforting behaviors. In Feinberg's (1977) study of female college students' responses to a "distressed" confederate, the need for help was manipulated by the confederate telling half the subjects that her boyfriend had dropped her the previous night (high need), and the other half that her boyfriend had dropped her the previous year (low need). As expected, subjects in the high need condition produced both a greater number of comforting behaviors and more sensitive comforting behaviors than subjects in the low need condition. Complementing this finding, Yarrow and Waxler (1976) showed that children who frequently received aggression from other children were more likely to be comforted by peers, provided that the aggression-receiving children were not themselves overly aggressive. Taken together, these results suggest that sensitive comforting is more likely to occur when the distress cues of another are viewed as highly salient.

Perceived similarity with another has been found to increase empathic responsiveness (see Barton & Coke, 1981). Given the posited association between empathy and comforting, it has been hypothesized that distressed others whom an actor perceives as similar to himself or herself will be more likely to receive comfort than will dissimilar others. The results of two studies lend support to this notion: In a naturalistic observation study of preschoolers,

Strayer (1981) found that comforting acts were never directed toward adults, but only other children. Zahn-Waxler et al. (1982) employed a repeated-measures design, exposing six- and seven-year-old children to both an adult and a peer who feigned minor physical injuries. Children were generally found to respond more solicitously to the peer's injury than the adult's injury. The findings of these two studies are supplemented by other findings indicating that prosocial interactions occur more frequently between children of the same sex than between children of opposite sexes (Eisenberg-Berg & Hand, 1979).

To summarize, sensitive comforting activity is more likely to occur when the distressed other has a close relationship with the comfortor, exhibits salient signs of distress, and is perceived by the comforter as similar to himself or herself.

Features of the Distressful Situation

At least two studies have found that certain characteristics of the distressful situation affect comforting behavior. Grodman (1979) employed a research design similar to that developed by Feinberg (1977) to study the comforting reactions of female college students to a confederate feigning emotional distress. Utilizing Feinberg's high-need condition, Grodman manipulated the cost of providing comfort to the confederate by telling half the subjects that they would receive important feedback about their personalities only if they completed all the experimental questionnaires they had been given to work on (high-cost-condition). No promise of feedback was given to the subjects in the low-cost condition. Although the results of this study were quite complex, subjects generally provided more comfort to the confederate in the low-cost condition.

Burleson (1982a) found that children used fewer and less sensitive comforting strategies in responding to a hypothetical situation involving a peer who was angry with a liked teacher than in responding to three other hypothetical situations. Burleson suggested that the children may have devalued the peer's feelings, and therefore provided less comfort, due to the tendency of children to revere authority figures.

In general, all of the situational variables reviewed in this section (states of the speaker, characteristics of the distressed other, and features of the context) have been interpreted as affecting the *motivation* to provide comfort. Although this interpretation is certainly debatable, it is at least plausible given the existing results. Moreover, this is an interpretation that can be accommodated within a general model of the comforting process (see below).

ANTECEDENTS OF COMFORTING SKILL

What factors lead children to acquire both the predisposition to provide comfort and sophisticated comforting skills? Very few studies have examined this question directly. There is, however, a considerable literature concerned with the antecedents of prosocial behavior in general (see Brody & Shaffer, 1982). Best known, perhaps, is Hoffman's (1977a) work on the influence of "power-assertive" versus "inductive" parental styles. A power-assertive style attempts to bring a child's behavior into conformity with parental expectations through the direct exercise of authority, or through material rewards and punishments. In contrast, an inductive style seeks to produce conformity with parental expectations through the use of reasoning, particularly about how an act or its consequences will affect others. Hoffman's distinction between power-assertive and inductive parental styles is quite similar to Bernstein's (1975) distinction between position-centered and person-centered communication (for a detailed integration of Hoffman's and Bernstein's distinctions, see Applegate et al., in press).

Parents employing an inductive style of discipline have consistently been found to have more prosocial children than parents practicing a power-assertive disciplinary style (Dlugokinski & Firestone, 1974; Hoffman, 1975a; Hoffman & Saltzstein, 1967; Brody & Shaffer, 1982). Hoffman (1975b) argues convincingly that the causal connection implied by these correlational findings is most parsimoniously viewed as running from parent to child; that is, variations in parental practices produce individual differences in children's prosocial behavior. However, it is not entirely clear which specific features of an inductive style serve to promote increased altruistic activity by children (see Rushton, 1980, pp. 118-119), nor is the theoretical character of this posited causal link entirely clear.

As Rushton (1980, pp. 124-125) indicates, the fact that an inductive style by parents leads to increased altruistic behavior by children can be explained in terms of (1) operant conditioning (altruistic acts by children are positively reinforced by induction); (2) social or observational learning (induction offers models of altruistic behavior that children imitate); (3) enhanced cognitive development (by pointing out the consequences of acts from others' perspectives, induction serves to stimulate moral development and the growth of social cognitive skills); or (4) self-concept growth (induction builds a strong self-concept, which in turn leads to increased altruistic action). It should be obvious that these possibilities are neither exhaustive nor mutually exclusive. For example, it is likely that an inductive style both models appropriate behaviors (thereby teaching children behavioral and communicative strategies) and enhances social cognitive development (thereby equipping children with the means of recognizing and understanding others' needs).

All of this suggests that parents who employ an inductive style or person-centered communication should have children who are more likely to provide comfort to distressed others through more developed comforting skills. Though it is rather sparse, the available empirical evidence supports this view. Zahn-Waxler, Radke-Yarrow, and King (1979) found that young children whose parents practiced an inductive style were more likely to produce comforting acts in experimentally structured situations, and more frequently engaged in spontaneous acts of comforting in natural situaitons, than children whose parents typically employed a power-assertive style. Similar findings have been reported by other researchers (for example, Bryant & Crockenberg, 1980; Hoffman, 1963).

Delia, Burleson, and Kline (1979) investigated the impact of the person-centeredness of mothers' disciplinary strategies on the sensitivity of comforting strategies employed by their grade school children. Person-centeredness of maternal disciplinary strategies was moderately related (r = .50) to the quality of the children's comforting strategies. One year later, the quality of children's comforting strategies was reassessed; maternal communication continued to be predictive of these strategies (r = .42). These results are consistent with other research indicating that person-centered parental communication affects children's referential and persuasive communication skills (Bearison & Cassel, 1975; Jones, Delia, & Clark, 1981b).

Of particular interest in this latter study was the fact that maternal disciplinary strategies were found to be predictive of children's comforting strategies. The fact that two different functional forms of communication were found to be related suggests that modeling cannot be solely responsible for the link between parent-child behavior. Rather, as Applegate and Delia (1980) cogently argue, it would appear that person-centered maternal communication enhances the social cognitive development of children by making the perspectives of others salient, and by elaborating specific features of these perspectives (motives, feelings, intentions, needs, and so on). The child exposed to such implicit instruction is thus led to focus spontaneously on the feelings and needs of others in a variety of contexts.

Research on the antecedents of comforting skills has barely begun, and little can be concluded at this time. While it appears that parental communicative styles do affect children's comforting skills, the precise nature of this influence is not well understood.

CONCLUSION:
TOWARD A MODEL OF
COMFORTING BEHAVIOR

Comforting behavior has been found to be related to several individual-difference and situational variables. Do these relationships suggest anything about the general character of the comforting process? This section sketches a general model of the comforting process based on existing research findings. The model is intended both to organize existing findings and to suggest directions for future research. The model begins by differentiating between two broad classes of variables: those that affect the competence to provide sensitive comfort, and those that affect the motivation to provide comfort.

Competence-Mediating Variables

By inquiring about the determinants of the competence to provide comfort, we are attempting to determine what skills and abilities a speaker needs to possess if he or she is to comfort others effectively. Burleson (1981) suggests that speakers must have at least three distinct types of knowledge if they are to be regarded as possessing the ability to comfort distressed others. First, a potential comforter must be capable of acquiring knowledge about the feelings and psychological states of the distressed other. As

noted earlier, sensitive comforting presupposes an awareness of
the distressed other's affective condition. Social perception processes, including affect recognition, impression formation and
organization, and perspective taking, are the chief means through
which a potential comforter acquires knowledge about the affective states of others. Thus, a skilled comforter must possess
developed social perception skills through which the emotional
state of a distressed other can be recognized and appreciated. As
noted above, this view is supported by many studies showing
positive relationships between indices of social perception ability
and comforting skills.

Second, the skilled comforter must possess knowledge about
the topics on which he or she is going to speak. Obviously, the
topics of primary import to a comforter concern human emotion
and the circumstances or situations in which human emotion is
a particularly salient feature. This means that a skilled comforter,
in addition to having knowledge about the specific emotions of
a particular other, must also have some general understanding of
the dynamics of human motivation and emotion. So far, the relationship between a general understanding of emotional dynamics
and sensitive comforting has not been directly assessed (although
it is likely that assessments of social perception skills also tap
general knowledge about emotional dynamics). Whiteman's (1967)
task for measuring developmental differences in conceptions of
psychological causality is one promising method for assessing
general knowledge of human emotional dynamics. This task
assesses the extent to which children are capable of understanding others' actions in terms of psychodynamic adjustment
mechanisms such as projection, repression, displacement, and
denial. The preceding logic suggests that people having a better
understanding of these mechanisms should also be more skilled
comforters.

Third, the skilled comforter must possess some knowledge of
the linguistic and rhetorical resources through which comforting
intentions can be realized. In order to comfort another, it is not
enough merely to understand the other's psychological state
(listener knowledge) and the general character of human emotional
dynamics (topic knowledge). Rather, comforters must also possess
a repertoire of behavioral strategies and tactics through which
knowledge of the listener and knowledge of the topic can be integrated and effectively applied. Thus, individuals possessing a
rich repertoire of strategies should be more capable of flexible,

effective, and sensitive comforting than persons with a relatively sparse repertoire of strategies. Empirical support for this notion is given in studies finding positive relationships between assessments of the number and the sensitivity of strategies employed in responding to hypothetical situations (O'Keefe & Delia, 1979). In addition, some research (for example, Delia & Clark, 1977) indicates that although a speaker may have a clear understanding of another's perspective, communication cannot succeed unless the speaker has mastered strategies for conveying his or her intentions. Unfortunately, little is known about the factors leading speakers to develop a rich and sophisticated repertoire of communicative strategies.

To summarize, the competence to comfort others depends on the possession of listener knowledge, topic knowledge, and rhetorical knowledge. Thus, persons possessing more developed social perception skills, a better understanding of human emotional dynamics, and a broader repertoire of rhetorical strategies should have a greater ability to comfort distressed others.

Motivation-Mediating Variables

Possession of the competence to comfort others does not, of course, ensure that this competence will be exercised. If sensitive comforting is to occur, a speaker must also be motivated to reduce another's distress. Personality traits and situational variations can both be viewed as primarily affecting the motivation to comfort distressed others.

There are two types of personality traits that affect the motivation to provide comfort: those that affect the desire to provide comfort, and those that affect the willingness to provide comfort. Traits found to influence the desire to provide comfort include the salience of interpersonal values (Borden, 1979), prosocial orientation (Feinberg, 1977), and emotional empathy (Burleson, in press). The salience of interpersonal values and prosocial orientation may be viewed as *generally* operative in that their influence extends across variations in situations. On the other hand, a trait such as emotional empathy appears to function in a situationally dependent manner; an emotionally empathic disposition has been found to result in increased help to others primarily when distress cues are particularly noticeable or large (see Barnett, this volume; Mehrabian & Epstein, 1972). Thus it appears that some personality traits generally affect the desire to provide comfort, while other traits only affect this desire depending on cues in the situation. This notion can be tested by assessing the influence of different desire-mediating traits in low-need and high-need for help conditions (see Feinberg, 1977).

So far, no research has examined whether comforting behavior is influenced by personality traits that can be viewed as affecting the willingness to provide comfort. For example, it might be expected that the locus of control orientation would affect the willingness to engage in comforting acts. People with a high external locus of control presumably would believe that they could not have an effect on another's distress, and thus would be less willing to engage in comforting acts, while those with a high internal locus of control would be more willing to engage in such acts because of a general belief in their ability to influence events. There is some indirect support for this hypothesis; individuals with a high internal locus of control have been found to behave more prosocially (Fincham & Barling, 1978; Midlarsky & Midlarsky, 1976) and are more likely to engage in goal-oriented communicative behavior (Phares, 1965). In addition, it is possible that personality traits such as "unwillingness to communicate" (Burgoon, 1976) and "communication apprehension" (McCroskey, 1977) affect the willingness to provide comfort to distressed others.

Varied features of communicative situations can also be conceptualized as affecting the desire and/or willingness to comfort others. For example, Applegate's (1980a) finding that daycare teachers employed less sensitive comforting strategies when they were tired, upset, or frustrated can be viewed as indicating that these temporary states reduce the desire and/or willingness to provide comfort. Similarly, characteristics of the distressed other (for example, relationship with the speaker, apparent need for comfort, similarity to the speaker) and characteristics of the situation (for example, cost of helping or conflict between liked others) may also be viewed as influencing the desire and/or willingness to provide comfort. As noted earlier, situational influences may combine or interact with personality influences in determining specific levels of motivation to provide comfort. In addition, there is evidence that certain situational features influence displayed levels of social perception skill (Bearison, 1976). It is thus possible that situational variables may systematically influence contextually realizable levels of comforting competence.

Developing an understanding of comforting communication that embraces competence factors, motivational factors, and their interrelationships is a challenge that should keep researchers busy for some time. Future research should address the substantive gaps in our understanding of comforting communication and reflect an awareness of the particular methodological issues concerning data collection and coding presented here.

REFERENCES

Allen , R. R., & Brown, K. L. (Eds.). (1976). Developing communicative competence in children. Skokie, IL: National Textbook.

Alvy, K. T. (1973). The development of listener-adapted communications in gradeschool children from different social-class backgrounds. Genetic Psychology Monographs, 87, 33-104.

Applegate, J. L. (1978). Four investigations of the relationship between social cognitive development and person-centered regulative and interpersonal communication. Unpublished doctoral dissertation, University of Illinois at Urbana-Champaign.

Applegate, J. L. (1980a). Person- and position-centered communication in a day-care center. In N. K. Denzin (Ed.), Studies in symbolic interaction: Vol. 3. Greenwich, CT: JAI Press.

Applegate, J. L. (1980b). Adaptive communication in educational contexts: A study of teachers' communicative strategies. Communication Education, 29, 158-170.

Applegate, J. L. (1982). The impact of construct system development on communication and impression formation in persuasive contexts. Communication Monographs, 49, 277-289.

Applegate, J. L., Burke, J. A., Burleson, B. R., Delia, J. G., & Kline, S. L. (in press). Reflection-enhancing parental communication. In I. E. Sigel (Ed.), Parents' constructions of child development. Hillsdale, NJ: Lawrence Erlbaum.

Applegate, J. L., & Delia, J. G. (1980). Person-centered speech, psychological development and the contexts of language usage. In R. St. Clair & H. Giles (Eds.), The social and psychological contexts of language. Hillsdale, NJ: Lawrence Erlbaum.

Asher, S. R. (1979). Referential communication. In G. J. Whitehurst & B. J. Zimmerman (Eds.), The functions of language and cognition. New York: Academic Press.

Baldwin, T. L., McFarlane, P. T., & Garvey, C. J. (1971). Children's communication accuracy related to race and socioeconomic status. Child Development, 42, 345-357.

Barenboim, C. (1977). Developmental changes in the interpersonal cognitive system from middle childhood to adolescence. Child Development, 48, 1467-1474.

Bar-Tal, D., Raviv, A., & Goldberg, M. (1982). Helping behavior among preschool children: An observational study. Child Development, 53, 396-402.

Batson, D., & Coke, J. S. (1981). Empathy: A source of altruistic motivation for helping? In J. P. Rushton & R. M. Sorrentino (Eds.), Altruism and helping behavior: Social, personality, and developmental perspectives. Hillsdale, NJ: Lawrence Erlbaum.

Bearison, D. J. (1976). Intraindividual variation in the coordination of social perspectives. Social Behavior and Personality, 4, 309-314.

Bearison, D. J., & Cassel, T. Z. (1975). Cognitive decentration and social codes: Communicative effectiveness in young children from different family contexts. Developmental Psychology, 11, 29-36.

Berkowitz, L. (1968). Responsibility, reciprocity, and social distance in help giving. *Journal of Experimental Social Psychology, 4,* 46-63.

Bernstein, B. (1975). *Class, codes, and control: Theoretical studies towards a sociology of language* (rev. ed.). New York: Schocken Books.

Borden, A. W. (1979). *An investigation of the relationships among indices of social cognition, motivation, and communicative performance.* Unpublished doctoral dissertation, University of Illinois at Urbana-Champaign.

Brody, G. H., & Shaffer, D. R. (1982). Contributions of parents and peers to children's moral socialization. *Developmental Review, 2,* 31-75.

Bryant, B. K., & Crockenberg, S. B. (1980). Correlates and dimensions of prosocial behavior: A study of female siblings with their mothers. *Child Development, 51,* 529-544.

Burgoon, J. K. (1976). The unwillingness-to-communicate scale: Development and validation. *Communication Monographs, 43,* 60-69.

Burleson, B. R. (1978). *Relationally oriented construct system content and messages directed to an affectively distressed listener: Two exploratory studies.* Paper presented at the annual convention of the Speech Communication Association.

Burleson, B. R. (1980). The development of interpersonal reasoning: An analysis of message strategy justifications. *Journal of the American Forensic Association. 17,* 102-110.

Burleson, B. R. (1981). *The influence of age, construct system developments, and affective perspective-taking skills on the development of comforting message strategies.* Paper presented at the annual convention of the International Communication Association.

Burleson, B. R. (1982a). The development of comforting communication skills in childhood and adolescence. *Child Development, 53,* 1578-1588.

Burleson, B. R. (1982b). The affective perspective-taking process: A test of Turiel's role-taking model. In M. Burgoon (Ed.), *Communication Yearbook VI.* Beverly Hills, CA: Sage.

Burleson, B. R. (in press). Effects of social cognition and empathic motivation on adults' comforting strategies. *Human Communication Research.*

Burleson, B. R., Applegate, J. L., & Neuwirth, C. M. (1981). Is cognitive complexity loquacity? A reply to Powers, Jordan, and Street. *Human Communication Research, 7,* 212-225.

Burleson, B. R., & Delia, J. G. (1983). *Adaptive communication skills in childhood: A unitary construct?* Paper presented at the biennial meeting of the Society for Research in Child Development.

Clark, R. A., & Delia, J. G. (1979). Topoi and rhetorical competence. *Quarterly Journal of Speech, 65,* 187-206.

Cowen, E. L., Pederson, P. Babigian, J., Izzo, L. D., & Tost, M. A. (1973). Long-term follow-up of early detected vulnerable children. *Journal of Consulting and Clinical Psychology, 41,* 438-446.

Damon, W. (1977). *The social world of the child.* San Francisco: Jossey-Bass.

Delia, J. G., Burleson, B. R., & Kline, S. L. (1979). *Person-centered parental communication and the development of social cognitive and communicative abilities: A preliminary longitudinal analysis.* Paper presented at the annual convention of the Central States Speech Association.

Delia, J. G., Burleson, B. R., & Kline, S. L. (in press). Developmental changes in children's and adolescents' interpersonal impressions. *Journal of Genetic Psychology.*

Delia, J. G., & Clark, R. A. (1977). Cognitive complexity, social perception, and the development of listener-adapted communication in six-, eight-, ten-, and twelve-year-old boys. *Communication Monographs, 44*, 326-345.

Delia, J. G., Clark, R. A., & Switzer, D. E. (1974). Cognitive complexity and impression formation in informal social interaction. *Speech Monographs, 41*, 299-308.

Delia, J. G., Kline, S. L., & Burleson, B. R. (1979). The development of persuasive communication strategies in kindergarteners through twelfth-graders. *Communication Monographs, 46*, 241-256.

Delia, J. G., & O'Keefe, B. J. (1979). Constructivism: The development of communication. In E. Wartella (Ed.), *Children communicating*. Beverly Hills, CA: Sage.

Delia, J. G., O'Keefe, B. J., & O'Keefe, D. J. (1982). The constructivist approach to communication. In F.E.X. Dance (Ed.), *Human communication theory*. New York: Harper & Row.

Dlugokinski, E., & Firestone, I. J. (1973). Congruence among four methods of measuring other-centeredness. *Child Development, 44*, 304-308.

Dlugokinski, E. L., & Firestone, I. J. (1974). Other-centeredness and susceptibility to charitable appeals: The effects of perceived discipline. *Developmental Psychology, 10*, 21-28.

Eisenberg-Berg, N., & Hand, M. (1979). The relationship of preschoolers' reasoning about prosocial moral conflicts to prosocial behavior. *Child Development, 50*, 356-363.

Eisenberg-Berg, N., & Lennon, R. (1980). Altruism and the assessment of empathy in the preschool years. *Child Development, 51*, 552-557.

Enright, R. D., & Lapsley, D. K. (1980). Social role-taking: A review of the constructs, measures, and measurement properties. *Review of Educational Research, 50*, 647-674.

Feinberg, H. K. (1977). *Anatomy of a helping situation: Some personality and situational determinants of helping in a conflict situation involving another's psychological distress*. Unpublished doctoral dissertaiton, University of Massachusetts, Amherst.

Feshbach, N. D. (1978). Studies of empathic behavior in children. In B. A. Maher (Ed.), *Progress in experimental personality research: Vol. 8*. New York: Academic Press.

Feshbach, N. D., & Roe, K. (1968). Empathy in six- and seven-year-olds. *Child Development, 39*, 133-145.

Fincham, F., & Barling, J. (1978). Locus of control and generosity in learning disabled, normal achieving, and gifted children. *Child Development, 49*, 530-533.

Flavell, J. H. (1968). *The development of role taking and communication skills in children*. New York: John Wiley.

Ford, M. E. (1979). The construct validity of egocentricism. *Psychological Bulletin, 86*, 1169-1188.

Gergen, K. J., Gergen, M. M., & Meter, K. (1972). Individual orientations to prosocial behavior. *Journal of Social Issues, 28*, 105-130.

Glucksberg, S., Krauss, R., & Higgins, E. T. (1975). The development of referential communication skills. In F. D. Horowitz (Ed.), *Review of child development research: Vol. 4*. Chicago: University of Chicago Press.

Goffman, E. (1967). *Interaction ritual: Essays on face-to-face behavior*. New York: Anchor Books.

Grodman, S. M. (1979). *The role of personality and situational variables in respond-*
ing to and helping an individual in psychological distress. Unpublished doctoral dissertation, University of Massachusetts, Amherst.

Hale, C. L. (1980). Cognitive complexity-simplicity as a determinant of communicative effectiveness. *Communication Monographs, 47,* 304-311.

Hale, C. L. (1982). An investigation of the relationship between cognitive complexity
and listener-adapted communication. *Central States Speech Journal, 33,*
339-344.

Hale, C. L., & Delia, J. G. (1976). Cognitive complexity and social perspective-taking.
Communication Monographs, 43, 195-203.

Higgins, E. T. (1976). Social class differences in verbal communication accuracy:
A question of "which question." *Psychological Bulletin, 83,* 695-714.

Hoffman, M. L. (1963). Parent discipline and the child's consideration for others.
Child Development, 34, 573-588.

Hoffman, M. L. (1975a). Altruistic behavior and the parent-child relationship. *Journal*
of Personality and Social Psychology, 31, 937-943.

Hoffman, M. L. (1975b). Moral internalization, parental power, and the nature of
parent-child interaction. *Developmental Psychology, 11,* 228-239.

Hoffman, M. L. (1977a). Moral internalization: Current theory and research. In L.
Berkowitz (Ed.), *Advances in experimental social psychology: Vol. 10.* New York:
Academic Press.

Hoffman, M. L. (1977b). Sex differences in empathy and related behaviors.
Psychological Bulletin, 84, 712-722.

Hoffman, J. L., & Saltzstein, H. (1967). Parent discipline and the child's moral
development. *Journal of Personality and Social Psychology, 5,* 45-57.

Hogan, R. (1969). Development of an empathy scale. *Journal of Consulting and*
Clinical Psychology, 33, 307-316.

Iannotti, R. J. (1981). *Prosocial behavior, perspective taking, and empathy in*
preschool children: An evaluation of naturalistic and structured settings. Paper
presented at the biennial meeting of the Society for Research in Child
Development.

Jones, J. L., Delia, J. G., & Clark, R. A. (1981a). *Socio-economic status and the*
developmental level of second- and seventh-grade children's persuasive
strategies. Paper presented at the annual convention of the International Communication Association.

Jones, J. L., Delia, J. G., & Clark, R. A. (1981b). *Person-centered parental com-*
munication and the development of communication in children. Paper presented
at the annual convention of the International Communication Association.

Kelly, G. A. (1955). *The psychology of personal constructs* (2 vols.). New York: W.
W. Norton.

Krauss, R. M., & Glucksberg, S. (1969). The development of communication: Competence as a function of age. *Child Development, 40,* 255-266.

Krebs, D. L. (1970). Altruism: An examination of the concept and a review of the
literature. *Psychological Bulletin, 73,* 258-302.

Krebs, D. L., & Russell, D. (1981). Role-taking and altruism: When you put yourself
in the shoes of another, will they take you to their owner's aid? In J. P. Rushton
& R. M. Sorrentino (Eds.), *Altruism and helping behavior: Social, Personality,*
and developmental perspectives. Hillsdale, NJ: Erlbaum.

Kurdek, L. A. (1977). Structural components and intellectual correlates of cognitive
perspective taking in first- through fourth-grade children. *Child Development,*
48, 1503-1511.

Kurdek, L. A. (1978). Perspective taking as the cognitive basis of chidren's moral development: A review of the literature. *Merrill-Palmer Quarterly, 24,* 3-28.

Ladd, G. W. (1981). Effectiveness of a social learning method for enhancing children's social interaction and peer acceptance. *Child Development, 52,* 171-178.

Loevinger, J., & Wessler, R. (1970). *Measuring ego development: Vol. 1.* San Francisco: Jossey-Bass.

Macaulay, J., & Berkowitz, L. (Eds.). (1970). *Altruism and helping behavior.* New York: Academic Press.

McCroskey, J. C. (1977). Oral communication apprehension: A summary of recent theory and research. *Human Communication Research, 4,* 78-96.

Mehrabian, A., & Epstein, N. (1972). A measure of emotional empathy. *Journal of Personality, 40,* 525-543.

Midlarsky, M., & Midlarsky, E. (1976). Status inconsistency, aggressive attitude, and helping behavior. *Journal of Personality, 44,* 371-391.

Murphy, L. B. (1937). *Social behavior and child personality.* New York: Columbia University Press.

Mussen, P., & Eisenberg-Berg, N. (1977). *Roots of caring, sharing, and helping: The development of prosocial behavior in children.* San Francisco: W. H. Freeman.

Nidorf, L. J., & Crockett, W. H. (1965). Cognitive complexity and the integration of conflicting information in written impressions. *Journal of Social Psychology, 66,* 165-169.

Oden, S. L., & Asher, S. R. (1977). Coaching children in social skills for friendship making. *Child Development, 48,* 495-506.

O'Keefe, B. J., & Delia, J. G. (1979). Construct comprehensiveness and cognitive complexity as predictors of the number and strategic adaptation of arguments and appeals in a persuasive message. *Communication Monographs, 46,* 231-240.

O'Keefe, B. J., & Delia, J. G. (1982). Impression formation and message production. In M. E. Roloff & C. R. Berger (Eds.), *Social cognition and communication.* Beverly Hills, CA: Sage.

O'Keefe, D. J., & Delia, J. G. (1981). Construct differentiation and the relationship of attitudes and behavioral intentions. *Communication Monographs, 48,* 146-157.

O'Keefe, D. J., & Sypher, H. E. (1981). Cognitive complexity measures and the relationship of cognitive complexity to communication: A critical review. *Human Communication Research, 8,* 72-92.

Payne, F. D. (1980). Children's prosocial conduct in structured situations and as viewed by others: Consistency, convergence and relationships with person variables. *Child Development, 51,* 1252-1259.

Phares, E. J. (1965). Internal-external locus of control as a determinant of amount of social influence exerted. *Journal of Personality and Social Psychology, 2,* 642-647.

Ritter, E. M. (1979). Social perspective-taking ability, cognitive complexity, and listener-adapted communication in early and late adolescence. *Communication Monographs, 46,* 40-51.

Roff, M., Sells, S. B., & Golden, M. W. (1972). *Social adjustment and personality development in children.* Minneapolis: University of Minnesota Press.

Rosenhan, D. L., Moore, B. S., & Underwood, B. (1976). The social psychology of moral behavior. In T. Lickona (Ed.), *Moral development and behavior.* New York: Holt, Rinehart & Winston.

Rubin, K. H. (1972). Relationship between egocentric communication and popularity among peers. *Developmental Psychology, 7,* 364.

Rubin, K. H. (1978). Role taking in childhood: Some methodological considerations. *Child Development, 49,* 428-433.

Rubin, K. H., & Schneider, F. W. (1973). The relationship between moral judgment, egocentrism, and altruistic behavior. *Child Development, 44,* 661-665.

Rushton, J. P. (1976). Socialization and the altruistic behavior of children. *Psychological Bulletin, 83,* 898-913.

Rushton, J. P. (1980). *Altruism, socialization, and society.* Englewood Cliffs, NJ: Prentice-Hall.

Samter, W. E. (1982). Personal communication.

Scarlett, H. E., Press, A. N., & Crockett, W. H. (1971). Children's descriptions of peers: A Wernerian developmental analysis. *Child Development, 42,* 439-453.

Selman, R. L. (1980). *The growth of interpersonal understanding: Developmental and clinical analyses.* New York: Academic Press.

Shantz, C. U. (1975). The development of social cognition. In E. M. Hetherington (Ed.), *Review of child development research: Vol. 5.* Chicago: University of Chicago Press.

Shantz, C. U. (1981). The role of role-taking in children's referential communication. In W. P. Dickson (Ed.), *Children's oral communication skills.* New York: Academic Press.

Spivack, G., Platt, J. J., & Shure, M. B. (1976). *The problem-solving approach to adjustment.* San Francisco: Jossey-Bass.

Staub, E. (1979). *Positive social behavior and morality: Vol. 2. Socialization and development.* New York: Academic Press.

Staub, E. (1980). Social and prosocial behavior: Personal and situational influences and their interactions. In E. Staub (Ed.), *Personality: Basic aspects and current research.* Englewood Cliffs, NJ: Prentice-Hall.

Strayer, F. F. (1981). The nature and organization of altruistic behavior among preschool children. In J. P. Rushton & R. M. Sorrentino (Eds.), *Altruism and helping behavior: Social, personality, and developmental perspectives.* Hillsdale, NJ: Erlbaum.

Strayer, J. (1980). A naturalistic study of empathic behaviors and their relation to affective states and perspective-taking skills in preschool children. *Child Development, 51,* 815-822.

Sypher, H. E., & Applegate, J. L. (1982). Cognitive complexity and verbal intelligence: Clarifying relationships. *Educational and Psychological Measurement, 49,* 537-543.

Underwood, B., & Moore, B. (1982). Perspective-taking and altruism. *Psychological Bulletin, 91,* 143-173.

Whiteman, M. (1967). Children's conceptions of psychological causality. *Child Development, 38,* 143-156.

Yarrow, M. R., & Waxler, C. Z. (1976). Dimensions and correlates of prosocial behavior in young children. *Child Development, 47,* 118-125.

Zahn-Waxler, C., Iannotti, R., & Chapman, M. (1982). Peers and prosocial development. In K. H. Rubin & H. S. Ross (Eds.), *Peer relationships and social skills in childhood.* New York: Springer-Verlag.

Zahn-Waxler, C., Radke-Yarrow, M., & Brady-Smith, J. (1977). Perspective-taking and prosocial behavior. *Developmental Psychology, 13,* 87-88.

Zahn-Waxler, C., Radke-Yarrow, M., & King, R. A. (1979). Child rearing and children's prosocial initiations toward victims of distress. *Child Development, 50,* 319-330.

PART II

DEVELOPMENTS IN CHILDHOOD: INFLUENCE PROCESSES

4

Verbal Social Reasoning and Observed Persuasion Strategies

David Forbes and David Lubin

Over the course of the last fifteen years, the study of social development has received increasing emphasis. In studies of children's social reasoning, numerous researchers have developed descriptions of how the understanding of social phenomena changes with increasing age. In studies of children's social behavior, investigators have succeeded in characterizing typical patterns of interaction at different ages, and in describing some of the changes in the complexity of social relations that accompany such development (Blasi, 1980; Chandler, 1977; Combs & Slaby, 1978; Flavell & Ross, 1980; Forbes, 1978; Hartup, 1979; Urbain & Kendall, 1980). Perhaps as a natural outgrowth of separate studies of social reasoning development and social

Authors' Note: *This research was supported by grants from the National Science Foundation (BNS-78-09119) and the National Institute of Mental Health (1-RO1-MH34723). We are grateful to Peter Van der Laan and Monika Schmidt for their help in coding the data for this project. Requests for reprints should be sent to David Forbes, Peer Interaction Project, 510 Larsen Hall, Harvard Graduate School of Education, Appian Way, Cambridge, MA 02138.*

behavior development, there has been a rise of interest in the project of relating the two areas. Recognizing that meaningful relations must exist between the development of organization in behavior and the development of regulatory cognitive processes, investigators have begun to implement research designs in which assessments of both social reasoning and social behavior are made. It is hoped that such studies will help us discover how these two areas of development are intertwined.

Perhaps because this line of synthetic inquiry is only in its beginning stages, initial efforts have provided results that are rather mixed, with studies alternately finding and failing to find significant relationships between reasoning and behavior variables (Blasi, 1980; Forbes, 1978; Shantz, 1975). More recent efforts have begun to highlight the complex problems one faces in relating reasoning and behavior through empirical investigation, and to outline the implications of this complexity for the design of empirical research (Blasi, 1980; Gerson & Damon, 1978; Higgins, 1980; Rubin, 1973; Turiel, 1979). Foremost among the issues raised in recent work is the need for great specificity in the definition of reasoning and behavior variables in studies seeking to relate these two areas.

The purpose of this chapter is to report a study of reasoning-behavior relations. The study was designed in light of current critical emphasis on the specificity of research variables. In the cognitive area, we focused on children's explanations regarding how the social behavior of one individual might bring about a change in the psychological state of a second individual who is the "target" of the behavior. The data were obtained from children's explanations of inferences they made about the psychological states of age-mates depicted in actual videotape interactions. In the domain of behavior, we focused on the strategies children used to persuade a playmate to cooperate during free play. That is, we looked at behavior during occasions when children were attempting to bring about a change in another's psychological state through their own socially directed actions.

By comparing the data on these two aspects of children's social development, we sought to explore the relationship between the ability to reason verbally about how psychological states are affected by others' behavior and the ability to influence another's psychological state through one's own behavior. We examined social interactions among children who were unacquainted with one another prior to participating in the study. We felt that under these circumstances, abstract verbal reasoning might play a major role in the regulation of behavior.

The development of increased complexity of one sort or another in children's persuasive strategies has been well documented by past research examining persuasive messages formulated for hypothetical audiences. Starting with Flavell's pioneering research over a decade ago (Flavell, Botkin, Fry, Wright, & Jarvis, 1968), investigators have described developmental change in persuasive communications in terms of the general role-taking abilities that are assumed to underlie them (see Blumenfield & Kinghorn, 1977; Findley & Humphrey, 1974; Nicholson, 1975; Piche, Rubin, & Michlin, 1978; Wood, Weinstein, & Parker, 1967).

These studies were not "reasoning-behavior" investigations per se, in that they did not employ separate assessments of social reasoning and persuasive behavior. Still, because they formulated changes in persuasive behaviors in terms of reasoning abilities that are implicit in persuasive behavior, the results were germane to our current interest. In broad terms, these analyses suggest that:

(1) Children's earliest persuasive communications take no account of the characteristics of the listener—that is, they give not justification as to why the listener in particular should change his or her behavior to conform with the speaker's desires.

(2) Early development in children's persuasive communications reflects an increase in matching between the form of the persuasive message and the characteristics of the listener. This early form of the adaptation of messages is limited, however, to external and/or stereotypic qualities of the listener.

(3) Further development in persuasive communications relies on attention to specific characteristics of the listener, including the listener's psychological attitudes toward or interest in the persuasive situation.

(4) The highest levels of development in persuasion skills involve anticipation of others' perspectives in reciprocal role-taking activity. This is manifested in the development of persuasive messages directed at objections that another might have to the speaker's request. Such messages are framed from the perspective of the listener.

Clark, Delia, and their colleagues (Clark & Delia, 1976; Delia & Clark, 1977; Delia, Kline, & Burleson, 1979; O'Keefe & Delia, 1979) have provided the only corpus of research that attempts to relate patterns of persuasive behavior directly to separately assessed social reasoning development. They examine children's "free descriptions" of playmates in the manner pioneered by Lively and Bromley (1973) and assess the cognitive complexity (that is, the number of traits used in a description) and the construct abstractness (the physical versus psychological nature of the traits used) of these descriptions. These social reasoning variables, which have themselves been shown to bear some relation to one another (see Crockett, 1965), are related to children's persuasive behavior in hypothetical persuasion tasks. Examples of persuasive behavior are classified according to the speaker's sensitivity to another's perspective, including the other's interests and possible counterarguments.

Data from the series of studies by Clark, Delia, and their colleagues support the relationship between persuasive behavior and both general cognitive complexity and abstractness in children's construct systems. Within a population of elementary school children, the authors found significant correlations between the levels of perspective taking underlying persuasion and independent assessments of both perspective-taking skills and cognitive complexity in the children's construct systems. Cognitive complexity was a significant predictor of the number of different persuasive appeals in a college-age population. Construct abstractness was related to the sophistication of persuasion strategies in diverse adult populations, from mothers to daycare workers. Finally, in a recent developmental study (Delia et al., 1979), the authors found that cognitive complexity (that is, the number of constructs) was the best predictor of persuasive sophistication in early and middle childhood, while construct abstractness was the best predictor of sophistication in adolescence.

Applegate (1982) extended this general paradigm of investigation into a study of persuasion strategies in a real-life interaction task. His subjects were brought to a laboratory setting in dyads, with each individual being given the task of persuading another to give up a five-dollar bill that had been given to them both at the beginning of the task. Applegate's results indicate that the abstractness (but not the complexity) of interpersonal constructs is related to persuasive ability. This finding extends the work of Delia et al. into the domain of real-life interactions; it also supports the notion that construct abstractness, not cognitive complexity, is a better predictor of persuasive ability in older subjects.

These findings regarding the relation between the development of interpersonal constructs and of persuasive ability are provocative, since they suggest that it is possible to link increases in the sophistication of persuasive behavior to changes in the separately assessed development of reasoning skills. Unfortunately, the psychological details of how construct systems relate to structure in persuasive behavior are unclear. Clark and Delia (1976) suggest that greater differentiation in a child's construct system should promote increased sophistication in role taking, and hence in persuasive behavior. However, no developmental principle is advanced to relate their quantitative measure of cognitive complexity to their structural categories of persuasion strategies. A more detailed explanation of how cognitive complexity underlies persuasion is called for and should probably be based on a less quantitative scale of cognitive complexity. While the notion of construct abstractness is a more structural description of children's reasoning, the relation between this dichotomous variable and a six-level developmental classification of persuasion strategies also lacks a detailed explication.

Applegate's work makes a major contribution to this body of research by indicating that success in real-life situations of persuasive interaction can be predicted using the construct abstractness variable. Again, however, the structural psychological basis for this relationship is not explicated. Furthermore, the performance of Applegate's college-age subjects in a real-life persuasive situation was reported to have only a low correlation with analogous performance measures in the hypothetical persuasion task ($r = .31$ for number of strategies; $.37$ for level of strategy). It is puzzling to see strong correlations between reasoning measures and persuasive behavior in hypothetical and real-life contexts when the various examples of persuasive behavior were not strongly related in the two contexts.

In general, it would seem that work of this type has served to lend credence to the idea that persuasive behavior may come under the sway of social reasoning skills that can be operationalized and studied outside the interpersonal persuasion context itself. What must now be added to this general observation is a more detailed account of the forms of social reasoning that bear such a relationship to persuasion, and an explication of the relation between a particular social reasoning skill and a particular form of persuasive action during real-life interactions. The present study can be viewed as an attempt to address just those issues, generated by the history of research on children's persuasion, and by work on person perception skills, particularly those involving persuasion.

Instead of person perception, we chose to focus on a social reasoning ability that bears a clear relationship to the persuasion process—namely, children's ability to explain how one individual's behavior might affect another's psychological state or attitude. This approach allowed a clear formulation of the hypothesis addressed by our research—that verbal explanatory constructs are called into play during real-life interactions. Taking as our model Applegate's studies of real-life behavior, we examined children's naturally occurring persuasion strategies, extending this trend to the study of development in childhood. A behavioral situation was created wherein subjects were asked to interact with unfamiliar others, a situation in which general explanatory constructs were most likely to be utilized in particular interpersonal contacts. In order to render the conceptual basis of our expectations about the form of reasoning behavior relations as clear as possible, we have relied on systems for describing children's verbal explanations and their actual persuasion strategies which share a single set of underlying qualitative distinctions.

METHOD

Subjects in the study were 24 children from the Cambridge public school system in Massachusetts. Half of the children were five years of age (range 4.9 to 5.3) and half were seven years of age (range 6.9 to 7.3). The children were studied in laboratory play groups of six same-age children, half boys and half girls. Each group of subjects participated in interview assessments (described below) and in after school sessions, each of which lasted for one hour. The entire series for each group consisted of twelve sessions over a period of three weeks.

Subjects were administered a social reasoning test at the conclusion of the series of twelve after school sessions. This assessment consisted of a structured interview based on a seven-minute videotape record of social interaction among three children who were not subjects in the study. The children were six years of age (that is, similar in age to both the five- and seven-year-old subjects). The interaction depicted on the videotape included three episodes in which one child was attempting to persuade another about some aspect of the interaction. The videotaped stimulus was frozen following each exchange of social behavior, and subjects were asked to make inferences about the psychological state (that is,

TABLE 4.1 Levels of Scoring for Social Reasoning Assessment

1. Mechanistic Stereotype: The child posits a mechanistic, one-to-one relation between environmental events and individuals' psychological reactions. The child
   thinks that all persons will perceive a social behavior in the same fashion, will
   think the same thoughts about the actor, and will have the same feelings, as a
   result of experiencing the behavior.

2. Reaction Subjectivism: The simple mechanistic relation between environmental
   events and individual psychological reactions has been transformed via the recognition that reactions to such events may differ from person to person. Subjects
   at this level acknowledge that different persons may have a variety of feelings in
   reaction to social behavior, and may think various things about the individual
   who displays the behavior. Subjects at this level still possess a naive realism concerning the status of the environmental event, however. That is, they still regard
   their interpretation of the meaning of a social behavior as the "real one."

3. Constructive Subjectivism: The child recognizes that perceptions of the behavior
   of others, as well as reactions to that behavior, exist on the plane of subjectivity.
   Subjects no longer maintain that an act can have a status such as "friendly" independent of the observer, but recognize that individuals will have subjective
   perceptions of others' behavior as well as subjective reactions to the behavior
   as perceived.

the thoughts and feelings) of each of the children depicted, based
on the social interaction that had just transpired.

Following a subject's inference of a psychological state, inquiry
was made regarding why the child felt that this inference was correct. This inquiry included (1) a request for an explanatory statement linking the observed event with the inferred thought or feeling, along with probes designed to clarify this explanatory statement; (2) a proposition by the interviewer of an alternative feeling
that a character might have in reaction to the event, with a request
for the subject's opinion about the possibility of this alternative;
(3) a request for an explanatory statement concerning the child's
rejection or acceptance of the suggested alternative feeling; (4)
a proposition by the interviewer of an alternative way of thinking
about (that is, perceiving) the event; and (5) a request for an explanatory statement concerning the child's acceptance or rejection of this alternative.

The children's social reasoning was classified using a system
of three levels, each of which represented a qualitatively distinct
mode of thought regarding the relation between social behavior
directed toward an individual and psychological reactions occa-

sioned by those behaviors. The system of three levels followed a logically ordered hierarchy, with each level representing a transformation of the person-environment relation that formed the basis of the previous level.

Table 4.1 depicts in graphic form the distinctions that separate reasoning at levels one, two, and three. The level one child views behavior as impinging mechanically on the psychological states of those who are subject to it. He or she maintains that changes in psychological states as the result of another's behavior are stereotypic in character—that is, the same for all individuals. In our social reasoning interview, this type of reasoning is perhaps best seen as the manifestation of egocentrism, which makes it difficult for a child to conceive of others perceiving or reacting to an event differently than he or she would. Following is an example of reasoning at level one:

DO YOU THINK THAT SEAN WILL GIVE BOBBY THE TRUCK?

Yes, because he said "please can I use it."

HOW COME THAT WILL WORK?

'cause when you ask nicely, they're going to give it to you.

COULD SEAN STILL NOT GIVE BOBBY THE TRUCK ANYWAY?

No, because he said "please, pretty please."

Reasoning at level two reflects conceptual development away from a purely egocentric viewpoint, with recognition that others may react to an event in a variety of ways, based on their individual subjectivity. However, the limits of level two conception are present in the form of a "naive realism" regarding social behavior. That is, observable behavior is still held to have some evaluative status independent of the observer and his or her reactive process. Level two children will insist, for example, that a particular act is "friendly" independent of the perception of the one at whom the behavior might be directed. An example of this type of reasoning follows:

HOW DOES SEAN FEEL ABOUT WHAT BOBBY JUST SAID?

I think that he will be happy.

COULD HE BE FEELING ANY OTHER WAY? COULD HE BE MAD?

Maybe, if he doesn't like people to help him.

COULD SEAN BE THINKING THAT BOBBY ISN'T FRIENDLY?

Bobby *is* being friendly. Maybe Sean doesn't like friendly kids.

At level three, the child recognizes the role of subjectivity in both perceptions of and reactions to the behavior of others. Level three subjects no longer maintain that a behavior can be "friendly" independent of the observer. They recognize that individuals have a subjective perception of others' behavior, as well as a subjective reaction. An example of reasoning at level three is seen in the following:

SO SEAN COULD BE MAD ABOUT BOBBY COMING OVER TO HELP. HOW COULD THAT BE?

Well, maybe he might think that Bobby wasn't really being friendly. Maybe he would think that Bobby wanted to start a fight again.

HOW COULD ASKING TO HELP NOT BE FRIENDLY?

It could be friendly to some people but maybe not . . . maybe Sean wouldn't think it was.

Children were assigned to the level that corresponded to their highest-level response during the interview. Assessment of the level of any individual response was made by examining the sequence of a child's responses to the ordered set of questions employed as probes following each inference of a psychological reaction.

Reliability for the coding of children's social reasoning was computed based on the independent coding of twelve protocols from the sample data. Cohen's (1968) weighted Kappa was used to evaluate the level of agreement in the data, adjusted for chance agreement. A Kappa of .96 was obtained, with a lower .05 confidence interval boundary of .87.

All laboratory sessions were videotaped for the purpose of gathering behavioral data on children's persuasive behavior. A system was used that incorporated three camera views with individual wireless microphones, so that each child's verbalizations were individually recorded. Only behavior during the first four sessions was analyzed for the purpose of the present study. It has been suggested that young children's social relationships with their peers do not stabilize for four to six days after initial contact (McGrew, 1972). Our decision to study the children's behavior prior to stabilization was based on the assumption that generalized social reasoning abilities like the one being assessed are most influential on behavior in situations where familiarity with the specific characteristics of another person is limited.

The selection of persuasion episodes for later coding was accomplished via "event" sampling (Altmann, 1974), where coders indicated chronological start and stop points for all interactions that fit the following definition:

> A persuasion sequence begins when one child makes a verbalization which calls for action (or cessation of action) on the part of a second child, where this second child fails to perform (or cease) the action, and the first child makes a second reference to the requested action. The first act in the persuasion sequence (that is, the start point) is considered to be the original request of the child. The persuasion sequence is considered to be ended when the second child performs the desired action or when the first child ceases to make reference to a request for action by the second child.

This definition excluded sequences of interaction where the request of one child for action by a second child was immediately met with compliance, since they were far too numerous and were felt to provide minimal information about the persuasion process. It also excluded sequences where more than one child was the target of a persuasive episode, since these sequences involved levels of complexity that were difficult to code in an unambiguous fashion, and were thus beyond the scope of the present study.

Reliability for the selection of persuasion sequences was computed on the basis of the independent coding of four hours of videotaped behavior. In all, 121 situations were identified by one coder, and 124 by a second coder. Of these nominations for codable events, all but fifteen were cases of agreement between the two coding efforts. This figure corresponds to a first-pass reliability of 88 percent. When the cases of disagreement were examined, eight involved data missed by one coder—that is, cases where the coder who had not noted the situation agreed that it was a codable one upon reviewing it. Such cases were not a cause of codes being assigned by one observer if the other coder felt no code could be assigned, but only a source of "lost data"—a reduction in the number of codable instances in the database. When these cases were ignored, the reliability of producing consensually agreed-upon instances of codable behavior was actually 95 percent.

Examples of children's persuasive behavior were scored using a system that focused on seven categories of behavior defined in observable, morphological terms. These seven categories were organized into three types of persuasion strategy, based on the form of social reasoning that was assumed to be required for using each type of persuasive behavior. Table 4.2 presents the seven behavior categories, organized into the three strategy types, along with definitions and examples of each.

The first three examples in this system were classified as "ritualistic" persuasion strategies because of assumptions about the manner in which a child encounters such behaviors in his or her social development. The use of entreaty was assumed to be a ritualistic strategy that the child encounters very early in life. Parental emphasis on saying "please," or using a pleading tone of request, is a practice acquired by the child in the context of this ritualistic early training. The de-escalation of a request was assumed to be a similar "magical" strategy to which children are exposed very early. The notion of compromise, "having just a little" or "doing it later," is common to early childhood socialization, and such a strategy may be acquired by the child as a ritualistic route toward gaining the compliance of others.

The use of norms or rules also seemed to be a ubiquitous feature of early parent-child discourse, functioning primarily for parents as a means for communicating regularities in the contingent interpersonal environment, and as a means for rendering moot any objections a child might raise about parental discipline or behavior in general. All three of these persuasion strategies were thought to be good candidates for strategies that would be employed by a child who approached interpersonal persuasion with a repertoire of mechanistic, ritualized behaviors.

The next three examples of behavior were considered to be "affect-oriented" because each seemed to be addressing the issue of how one might feel about the suggested activity while attempting to communicate the self's conviction that a positive affect was appropriate. In cases of "appeals to history," the child will try to communicate to another that he or she liked the activity in the past, and hence that he or she should like it again. "Testimonial" behavior appears to rest on the power of group reference and the notion that a positive attitude on the part of others will serve to orient the listener in a positive direction. "Effusion of affect" seems to be a "hard sell" orientation to the subjective reactions of another involving attempts to submerge the other's possible negative feel-

TABLE 4.2   Coding Categories for Children's Persuasive Strategies

Type One: "Ritualistic Persuasion." Ritualistic persuasion is assumed to be a reflection of social reasoning at level one, in which a child assumes that mechanistic relationships exist between external events and internal, psychological reactions of others to those events. The types of persuasive behavior included in this group are as follows:

(a) Entreaty. Subject makes reference to desired action on the part of the other using polite form of request or pleading tone of voice.

Example:

Child A: "Can we play with the dolls now?"
Child B: "There's not enough room."
Child A: "Oh, *please*, I want to play with them."

(b) Deescalation of Request. Subject restates original request in modified form, imposing a limitation on the degree or duration of cooperation requested from the other.

Example:

Child A: "Can we have dinner for the dolls now?"
Child B: "No, they're not hungry."
Child A: "Well, they could have a little snack."

(c) Appeal to Norms. Subject makes reference to desired action on the part of other with allusion to a social rule (reasonable or idiosyncratic in nature) that prescribes desired action or proscribes action the cessation of which is desired.

Example:

Child A: "I'm going to build a castle over here with you."
Child B: "No, it's in the way."
Child A: But you *can't have* all the sand."

Type Two: Persuasive Orientation to Affect.

(a) Appeals to History. Subject makes reference to past situations in which the other acted (or refrained from acting) in the manner the subject currently desires.

Example:

Child A: "Let's build a wall here."
Child B: "No."
Child A: "We could do it just like we did yesterday."

(b) Testimonial. Subject makes reference to other individuals' positive attitudes toward the desired behavior, or negative attitudes toward the undesired behavior.

Example:

Child A: "We should dig it out at the bottom."
Child B: (ignores)
Child A: "Luke dug one out at the bottom before, and Kevin said it was the biggest one in the world."

(continued)

TABLE 4.2 Continued

(c) Effusion of Affect. Subject presents suggestion with phrasing or tone of voice that communicates excitement or positive affect toward the idea.

Example:

Child A: "We're gonna build a castle right here, and it's gonna be a good one, and you're gonna like it."

Type Three: Persuasive Orientation to Construal.

(a) Clarification of Referent. Subject provides other with factual or quasi-factual information concerning the nature of the desired (or undesired) activity.

Example:

Child A: "We could build this bridge with big blocks."
Child B: "But there won't be enough room."
Child A: "When the blocks go sideways, they can all fit."

(b) Clarification of Intent. Subject provides other with information concerning the reason for requesting an action, with reference to the basis of the request in subject's friendly intentions toward the other, or subject's awareness of unfriendly intentions of other toward the self.

Example:

Child A: "Can I have some men with the wings?"
Child B: "No, they are all mine."
Child A: "I'm not gonna keep them for always—we could do it together."

ings about a suggested activity beneath a contagion of enthusiasm on the part of the speaker. In all three cases, these persuasive strategies appear to focus primarily on how the other is feeling, and to provide a variety of incentives for positive feelings.

The final two examples were classified as "construal-oriented" strategies because both seemed to address the reluctance of others to comply because of a different construal of the characteristics of a suggestion or of the intent with which it was made. "Clarification of referent" seeks to expand another's perception of the possibilities opened up by the suggested activity, presumably based on the conviction that a more complete understanding of these possibilities will lead the other to comply. "Clarification of intent"specifically addresses the other as a construing individual who is making inferences about the motives of the speaker in offering a suggestion. This strategy seeks to provide data that will shape the other's inferences about or construal of the speaker as an intentional being. In both cases, the emphasis is on altering the other's sense of *what is* in the social world.

In order to code these behaviors, each preselected persuasion episode was viewed using a "focal individual" technique (Altmann, 1974) in which the actions of each individual were followed continously. The behavior of that individual and of those with whom he or she interacted was coded in sequence. Multiple passes through the videotape records were made until all relevant interactions between all members of the play group had been coded.

The reliability of coding the behavior categories was computed using the independent blind coding of our assistants, each of whom coded ten persuasion strategies and then recoded the strategies originally coded by each of his or her three colleagues. Thus a total of thirty reliability observations for each of four coders, or 120 paired observations in all, constituted the reliability data. Because of the disparity in the total number of behavior types observed in the various persuasion categories, sequences to be coded for reliability were not chosen at random, but were preselected from the records of the original coder in order to provide a representation of each behavior type in the corpus of behaviors to be double-coded. Cohen's (1968) weighted kappa was again used to assess the level of reliability in this coding effort. A kappa of .76 was obtained, with a lower .05 confidence interval boundary of .68.

RESULTS

Social Reasoning Performance

In order to examine the relationship between our reasoning assessment and subject age, we undertook a chi-square analysis of the reasoning scores in an age-by-performance level (2 x 3) design. The result of this analysis approached but did not reach significance ($\chi^2 = 5.5$, $p < .07$). The magnitude of individual difference in the rate of development for the age range sampled was thus greater than the degree of difference that could be normatively related to chronological age. This may suggest that the period from five to seven is one of great variability regarding how a child thinks about the psychological states of others, and that some children may show precocious sophistication in this respect, while others develop at a more moderate pace.

Persuasion Strategies

In order to assess the evidence for an age-related process of behavioral development in terms of our three types of persuasion strategy, we performed a chi-square analysis of an age-by-strategy type (2 x 3) matrix.[1] Our expectation of a positive relationship between age and level of persuasion behavior was not supported by a significant chi-square statistic. This implies that our levels of persuasive behavior do not represent an age-related description. It also indicates that individual variation in the behavioral patterns of five- and seven-year-olds is greater than any age-related pattern of normative development that might exist with respect to our persuasion behaviors.

Regarding the evidence for hierarchic organization among the levels of persuasion behavior, we asked, for each subject, whether or not the appearance of behavior termed "construal-oriented" was always accompanied by the presence of behavior labeled "affect-oriented" and "ritualistic," and whether the appearance of behavior lableled "affect-oriented" was always accompanied by behavior labeled "ritualistic." In all, 23 of 24 cases conformed to a hierarchic pattern, with "construal-oriented" behavior always accompanied by "ritualistic" and "affect-oriented" behavior, and "affect oriented" behavior always accompanied by "ritualistic" behavior. Guttman scale analysis of these data yielded coefficients of scalability (.85) and reproducibility (.97) which suggest strong support for the hierarchic organization of persuasion strategy types. This finding lends credence to the notion that our classification of persuasion strategies is a valid description of development, despite a lack of relation between age and level of persuasion strategy.

Fisher exact probability procedures were utilized to explore statistically the hypothesis that the use of a particular persuasion strategy at a given level was related to one's reasoning at the level we assumed to be related to that persuasion strategy. The overall relationship of reasoning to behavior as we conceived it, as well as the relation of specific reasoning abilities to the use of specific persuasion strategies, was strongly supported by these analyses. The overall probability that subjects' reasoning levels were independent of their observed persuasion strategies was < .001. The probability that construal-oriented behavior occurs independent of reasoning at level three on the interview task is < .05; the probability that affect-oriented persuasion behavior occurs independent of reasoning at level two on the interview task is < .01; and the probability that only ritualistic behavior occurs independent of reasoning at level one on the interview task is also < .01.

CONCLUSIONS

It appears that our strategy of coordinated specification of reasoning and behavior variables, and our careful selection of a behavioral phenomenon and context of behavior observation, have been productive. The picture that emerges from the study is a reasonably succinct one. First, we found age-related trends in children's social reasoning, suggesting that this skill, as we assessed it, is developmental in nature. Second, while a strong relationship between age and optimal level of persuasion behavior was not obtained, the near-perfect Guttman scaling of persuasion behaviors across the three levels for all subjects supports the notion that behavior characterized by the use of persuasion strategies is also developmental in nature. More specifically, it supports the notion that the acquisition of persuasion strategies proceeds as a process of "layer-cake" development, or what Werner (1957) has called "hierarchic stratification." Finally, strong association was found between performance in our reasoning interview and patterns in children's use of persuasion strategies. This supports our suggestion that the two phenomena studied are indeed related.

Our findings regarding development in children's social reasoning ability per se are not completely novel. Clearly, the movement from level one to level two in our categories of social reasoning is movement away from stereotypic, egocentrically formulated conceptions of others toward decentration and a consequent orientation to individual differences—both in the construction of and reaction to events in the social world. This path of development is widely documented in the social reasoning literature on topics as diverse as children's spontaneous descriptions of others, their reasoning regarding social norms and rules, and their attribution of motives behind actions (compare Chandler, 1977; Forbes, 1978; Shantz, 1975). The movement from level two to level three in the reasoning description marks a transition from conceiving of others as passive, albeit subjective, receptors of experiences toward conceiving of others as active, intentional constructors of their social world. This process has also been documented in past work. Notable in this respect is Broughton's (1978) work on children's conceptions of the nature and workings of the mind, and about the nature of reality.

Our findings expand on the general thrust of efforts by past students of social reasoning in that we document the effect of these general developmental trends in social understanding on one very specific aspect of interpersonal relationships—the occasioning in others of psychological states as the result of an actor's social behavior. Several recent investigators have shown empirically that one may find considerable "decalage" in abstract dimensions of development across specific areas of content (compare Rubin, 1973). Reacting to this, others have called for a movement toward the use of social reasoning measures that are more specific in the thrust of their assessment (see Applegate and Delia, 1980). If development does not manifest itself consistently in broad conceptual dimensions that cross specific content areas, then more focused study of development vis-à-vis important areas of content in social understanding is necessary. We believe that the conceptual content of our reasoning assessments is such an area.

Our data would also seem to indicate a somewhat earlier sophistication of development, in broad conceptual terms, than that reported by other investigators. The appearance of sensitivity to others' intentional attempts at construing experience is formally akin to "thinking about thinking," an ability that Flavell (1978) places within the accomplishments of middle to late childhood. It also bears resemblance to "S thinks about O thinking about S," a reciprocal role-taking activity that has also been viewed as a developmental accomplishment of late childhood (Selman, 1971; Selman & Byrne, 1974). Our utilization of a video-based stimulus involving naturalistic interaction between children in a peer situation suggests that the development of sophisticated social understanding may take place at an earlier age than that suggested by past work. This finding is consistent with those of other workers (for example, Bearison & Gass, 1979; Shultz & Butkowski, 1977) whose use of video- or film-based stimuli in social reasoning tasks elicited more sophistication in young subjects than comparable studies using verbally presented hypothetical material.

Clearly, there is room for ingenuity in the design of tasks intended to explore the use of reasoning by young children. Especially in the study of social development, much greater attention to the interaction context of the assessment task is called for. It may often be the case that the goal of the child in the assessment interaction may differ from the goal of the experimenter. Indeed, the interaction between the child and the experimenter in the assessment situation should often be viewed as the primary datum. Eisenberg-Berg and Lennon's (1980) commentary on the nature of verbally based assessments of empathy, in which they discuss suspicions about children's overarching goal of pleasing the interviewer, is notable in this context.

Our findings regarding children's persuasive strategies provide information that is more novel in character. We have characterized persuasion strategies that do not take account of another's perspective in positive terms, by proposing that they proceed from a mechanistic belief about how one's actions affect another's behavior. We have observed a hierarchic organization of persuasive strategies within the repertoires of our subjects which supports the notion that primitive forms of persuasion strategy are mechanistic in nature. Garvey's (1975) study of children's requests and responses is one effort that supports this proposition. Her analysis of children's requests of one another, and their responses to requests, identifies several "meaning factors" which define the discourse domain in these interactions. Among these are references to a listener's obligation to perform a requested action (akin to our mechanistic "appeal to norms") and to the future goal of the speaker which underlies a request (akin to our construal-oriented "clarification of intent"). Garvey reports that utterances referring to obligation were made predominantly by her younger group (mean age four years), while reference to intent were made predominantly by her older group (mean age five years).

We have also expanded the distinction between considering and not considering the perspective of another to include two bases upon which another's perspective may differ from that of the speaker: a basis of construal or understanding of the persuasive request, and a basis of affect or attitude toward the request. Again, our finding of hierarchic organization in children's persuasive repertoires suggests that this distinction may be a valuable one for identifying levels of sophistication in the use of persuasive strategies. Such a distinction makes it possible to move beyond descriptions of development that focus on the "explicitness" of perspective taking in persuasion, or on the "number of features" that are addressed in persuasion, toward a description based on the psychological demands of a persuasion situation. More empirical work will be necessary to confirm the validity of this distinction as a descriptor of development in persuasive ability.

More important than either the interview performance data or the peer behavior data are our findings regarding the relation between these two phenomena. It appears from our results that there are indeed strong ties between the structure of subjects' explanations regarding inferences about others' psychological states, and the types of behavior used in interpersonal persuasion situations. This finding, the pursuit of which was the major goal of the study, deserves extended commentary as we close our discussion.

We should begin by noting that we feel the dichotomy between reasoning and behavior, when made in a global fashion, to be a misdirected distinction. It seems clear that the verbal performance of children in "reasoning" assessments is certainly a type of behavior—typically communicative behavior vis-à-vis an investigator. It also seems clear that social behavior—at least some of it—should be considered reasoning in action. In this light, we would propose that studies of reasoning-behavior relations involve a process of: (1) developing a theory about how a mental faculty might manifest itself in a real-life interaction context, with the recognition that such a manifestation would be relatively opaque and imbedded in a matrix of other deterministic factors; (2) developing experimental assessments aimed at eliciting selected behaviors that are assumed to bear a more transparent relationship to these mental faculties; and (3) investigating whether an association exists between patterns of performance in an experimental context and patterns of natural behavior, with the assumption that such an association would support the theoretical claims on which the study is based.

A natural outcome of this redefined version of reasoning-behavior studies is the recognition that there is no single paradigm for such research. The use of any particular experimental assessment of social cognition, and the focus on any particular feature of natural social behavior, will automatically place a given study within a constrained theoretical universe. The results of any particular study will then be pertinent only to the questions that lie within that universe.

Considered in this way, the study reported here can be said to proceed from three propositions: (1) that children utilize a structured understanding of others as one of the means for regulating their behavior in persuasion situations with unfamiliar others; (2) that one relatively direct means of inferring the nature of such structures for a given individual is to examine his or her verbal explanations of inferences about others' psychological states; and

(3) that another, albeit less direct, manifestation of these structures appears in certain discernible patterns of persuasive behavior, provided such behavior is categorized in a manner that is consistent with the distinctions made among various types of verbal explanation behavior.

Our finding of a strong association between the verbal explanations of our subjects and patterns in their persuasive behavior with unfamiliar peers suggests that we have been correct in assuming that these phenomena are related psychologically. One way of interpreting this is to suggest that our two measures are isomorphic manifestations of a single underlying "structure" of thought, which makes its appearance in the regulation of both verbal explanation behavior and peer persuasion behavior. However, this form of interpretation tends toward that vision of reasoning-behavior research that reifies the notion of reasoning apart from its various manifestations in behavior, and that supports the dichotomy that we have claimed to reject. Remembering that both our measures pertain to "reasoned behavior," we would suggest an alternative interpretation, namely that children involved in persuasion situations regulate their behavior in part via a process of reasoning that is verbal-explanatory in nature. This interpretation places our findings within the discourse domain of workers such as Luria (1961), who suggest that the regulation of behavior relies in some instances on an internalized form of verbalization. Our interpretation is also equivalent to suggesting, as does Damon (Gerson & Damon, 1978), that behavior regulation in some forms of social interaction is "deliberative" in nature and may involve a fairly conscious consideration of one's structured social knowlege.

The final message of our study is one about the importance of a priori conceptual analysis in the study of social development. The relationship that we found between our verbal explanation data and our social behavior data was far stronger than that usually found in this kind of work. We would maintain that this is because of our reliance on a single set of conceptual distinctions which could be used to categorize both verbal explanations of psychological inferences and actions in persuasion situations.

The number of reasoning/behavior relations still requiring conceptual analysis is quite large. We know little, for instance, about the conceptual underpinnings of understanding relations between expressive behavior and psychological states. Understanding social status and its relation to social interaction is equally unexplored from the standpoint of systematic analysis. The nature of group dynamics is another area open for analytic work, as is the role of self-presentaion in social interaction. In all these areas, we need to know more about what children need to know in order to use the sophisticated reasoning strategies that lie beneath the surface of mature social interaction.

The notion that faculties of reasoning play a role in the regulation of behavior is axiomatic for all who view behavior as a planned, goal-oriented phenomenon. The study of which reasoning skills contribute to the regulation of behavior in which contexts, and in what manner, will continue to be an important and challenging task for developmental psychology.

NOTE

1. All final analyses of behavior data reported here are based on collapsed frequencies, with individual behavior categories under each of the three "strategy types" combined. This procedure included the dismissal of counts with one category showing almost no frequency for the entire sample, as well as dismissal of one "type three" behavior which was used by all subjects and hence was found not to be a reflection of level three reasoning as originally assumed. For complete details of these preliminary data-handling procedures, see Forbes (1981).

REFERENCES

Altmann, J. (1974). Observational study of behavior: Sampling methods. *Behavior,* *49*, 227-267.

Applegate, J. L. (1982). The impact of construct system development on communication and impression formation in persuasive contexts. *Communication Monographs, 49*, 277-289.

Applegate, J. L., & Delia, J. G. (1980). Person-centered speech, psychological development, and the contexts of language usage. In R. St. Clair & H. Giles (Eds.), *Social and psychological contexts of language.* Hillsdale, NJ: Lawrence Erlbaum.

Bearison, D., & Gass, S. (1979). Hypothetical and practical reasoning: Children's persuasive appeals in different social contexts. *Child Development, 50*, 901-903.

Blasi, A. (1980). Bridging moral cognition and moral action: A critical review of the literature. *Psychological Bulletin, 88*, 1-45.

Blumenfield, P. C., & Kinghorn, S. N. (1977). *Influence of role perception on persuasion.* Paper presented at the biennial meeting of the Society for Research in Child Development.

Broughton, J. (1978). Development of concepts of self, mind, reality and knowledge. In W. Damon (Ed.), *New directions in child development: Vol. 1. Social cognition.* San Francisco: Jossey Bass.

Chandler, M. J. (1977). Social cognition: A selective review of current research. In W. F. Overton & J. M. Gallagher (Eds.), *Knowledge and development: Vol. 1. Advances in research and theory.* New York: Plenum Press.

Clark, R. A., & Delia, J. G. (1976). The development of functional persuasive skills in childhood and early adolescence. *Child Development, 47,* 1008-1014.

Cohen, J. (1968). Weighted Kappa: Nominal scale agreement with provision for scaled disagreement or partial credit. *Psychological Bulletin, 70,* 213-220.

Combs, M. L., & Slaby, D. A. (1978). Social skills training with children. In B. Lahey & A. Kazdin (Eds.) *Advances in child clinical psychology: Vol. 1.* New York: Plenum Press.

Crockett, W. H. (1965). Cognitive complexity and impression formation. In B. A. Maher (Ed.), *Progress in experimental personality research: Vol. 2.* New York: Academic Press.

Delia, J. G., & Clark, R. A. (1977). Cognitive complexity, social perception, and the development of listener-adapted communication in six-, eight-, ten- and twelve-year-old boys. *Communication Monographs, 44,* 326-345.

Delia, J. G., Kline, S. L., & Burleson, B. R. (1979). The development of persuasive communication strategies in kindergartners through twelfth-graders. *Communication Monographs, 46,* 241-256.

Eisenberg-Berg, N., & Lennon, R. (1980). Altruism and the assessment of empathy in the preschool years. *Child Development, 51,* 552-557.

Findley, G. E., & Humphreys, C. A. (1974). Naive psychology and the development of persuasive appeals in girls. *Canadian Journal of Behavioral Science, 6,* 75-80.

Flavell, J. H. (1978). Metacognitive development. In J. M. Scandura & C. J. Brainerd (Eds.), *Structural-process theories of complex human behavior.* Leyden: Sythoff & Noordhoff.

Flavell, J. H., Botkin, P. T., Fry, C. L., Wright, J. W., & Jarvis, P. E. (1968). *The development of role-taking and communication skills in children.* New York: John Wiley.

Flavell, J. H., & Ross, L. (Eds.). (1980). *New directions in the study of social-cognitive development.* Cambridge: Cambridge University Press.

Forbes, D. L. (1978). Recent research on children's social cognition: A brief review. *New Directions for Child Development, 1,* 123-139.

Forbes, D. L. (1981). *Verbal social reasoning and observed persuasion strategies in children's peer interactions.* Unpublished doctoral dissertation, Clark University, Worcester, MA.

Garvey, K. (1975). Requests and responses in children's speech. *Journal of Child Language, 2,* 41-64.

Gerson, R. P., & Damon, W. (1978). Moral understanding and children's conduct. In W. Damon (Ed.), *New directions in child development: Vol. 1.* San Francisco: Jossey Bass.

Hartup, W. W. (1979). Levels of analysis in the study of social interaction: An historical perspective. In M. E. Lamb, S. J. Suomi, & G. R. Stephenson (Eds.), *Social interaction analysis.* Madison: University of Wisconsin Press.

Higgins, E. T. (1980). Role-taking and social judgement: Alternative developmental perspectives and processes. In J. H. Flavell & L. Ross (Eds.), *New directions in the study of social-cognitive development.* Cambridge: Cambridge University Press.

Lively, W. J., & Bromley, D. B. (1973). *Person perception in childhood and adolescence.* New York: John Wiley.

Luria, A. R. (1961). *The role of speech in the regulation of normal and abnormal behavior.* New York: Liveright.

McGrew, W. C. (1972). Aspects of social development in nursery school children, with an emphasis on introduction to the group. In N. Blurton-Jones (Ed.), *Ethological studies of child behavior.* Cambridge: Cambridge University Press.

Nicholson, J. L. (1975). *The development of role-taking abilities and sociolinguistic competence in three interpersonal communication domains among Caucasian, Black and Spanish-American fourth, fifth, and sixth grade children.* Unpublished doctoral dissertation, University of Illinois at Urbana-Champaign.

O'Keefe, B. J., & Delia, J. G. (1979). Construct comprehensiveness and cognitive complexity as predictors of the number and strategic adaptation of arguments and appeals in a persuasive message. *Communication Monographs, 46,* 231-240.

Piche, G. L., Rubin, D., & Michlin, M. (1978). Age and social class in children's use of persuasive appeals. *Child Development, 49,* 773-780.

Rubin, K. H. (1973) Egocentrism in childhood: A unitary concept? *Child Development, 44,* 102-110.

Selman, R. L. (1971). The relaton of role-taking to the development of moral judgement in children. *Child Development, 42,* 79-91.

Selman, R. L., & Byrne, D. F. (1974). A structural-developmental analysis of levels of role-taking in middle childhood. *Child Development, 45,* 803-806.

Shantz, C. U. (1975). The development of social cognition. In E. M. Hetherington (Ed.), *Review of child development research: Vol. 5.* Chicago: University of Chicago Press.

Shultz, T., & Butkowsky, I. (1977). Young children's use of the scheme of multiple sufficient casuality in real versus hypothetical situations. *Child Development, 48,* 464-470.

Turiel, E. (1979). Distinct conceptual and developmental domains: Social convention and morality. *Nebraska symposium on motivation, 1977: Vol. 25.* Lincoln: University of Nebraska Press.

Urbain, E. S., & Kendall, P. C. (1980). Review of social-cognitive problem-solving interventions with children. *Psychological Bulletin, 88,* 109-143.

Werner, H. (1957). *Comparative psychology of mental development.* New York: International Universities Press.

Wood, J. R., Weinstein, E. A., & Parker, R. (1967). Children's interpersonal tactics. *Sociological Inquiry, 37,* 129-138.

5

Children's Conversations Within a Conflict-of-Interest Situation

Sherri Oden, Valerie A. Wheeler,
and Sharon D. Herzberger

Among the major research interests of investigators of children's social development is the examination of processes critical to children's peer relationships. A number of studies have included comparisons of children who had positive peer relationships with children who had few or problematic peer relationships. In studies where children were differentiated on the basis of peer sociometric analyses, peer acceptance and friendship scores were correlated with a variety of measures of social behavior and social knowledge or judgment. Overall, the more socially accepted and sought-after peers were found to evidence greater participation in peer activities (Gottman, 1977), behave more cooperatively during activities (Hartup, 1978), and provide more support and help to peers (Coie, Dodge, & Coppotelli, 1982). Children who lacked peer acceptance and friends showed an opposite behavioral pattern. These children either interacted infrequently with their peers, or their interactions frequently included aggressive and disruptive behavior. Some tended to seek activity help but failed to reciprocate (Coie et al., 1982).

Research on children's peer relationships has found no clear patterns regarding the role of communication processes, although there is some evidence indicating a relationship between communication competence and peer status. For example, from interviews with children (Ladd & Oden, 1979) and observations of small groups in the classroom (Putallaz & Gottman, 1981), children's social judgment as to when it is appropriate to initiate an instruction or directive was found to be related to peer status. It is often assumed that communication skills are related to or based on the ability to understand the role or perspective of the listener (see Rubin & Everett, 1982, for a review). However, studies conducted with children and adults have usually been based on global conceptions of perspective taking (see Applegate, Burke, Burleson, Delia, & Kline, in press, for a review) and have not provided sufficient specificity on the types of communication strategies that are likely to be important in peer relationships.

Some studies of children's communication competence have focused on children's ability to select word or object referents appropriate for listeners lacking certain information. In general, more competent communicators were thought to have a better ability to discriminate between word or clue referents and/or to take into account the perspective of the listener. Gottman, Gonso, and Rasmussen (1975) used a referential word pair task with elementary school children and found some evidence that competence on one task, the dissimilar word-pair task, was related to having friends, but only at the third-grade level.

Tesch and Oden (1981) recently found results similar to Gottman et al. (1975), but again only for third graders. In this study, third- and sixth-grade children were also asked to provide single-word clues to help an imaginary listener select the correct word in the similar and dissimilar word pairs and in two additional word pair sets that contained more affective or expressive similar and dissimilar word pairs. Positive relationships were found between competence on both the dissimilar expressive word pairs (and the previously employed dissimilar word pairs) and peer sociometric ratings both for playing and conversing. However, these and other studies have found that relationships between measures of referential communication tasks and peer status account for a relatively small share of the overall variance (for example, Rubin, 1972; Krantz, 1982).

Many observational studies have included simple ratings of children's peer interactions, such as appropriate (or positive) verbal and inappropriate (or negative) verbal behaviors. This approach provides only a quantitative assessment of chjildren's social effectiveness and does not reveal the variation of communication skills. Gottman and Parkhurst (1979) and Putallaz and Gottman (1981) represent examples of research that has focused on the identification of effective communication strategies in peer interactions. For example, these studies revealed that, in general, popular children appeared to be more agreeable and able to head off or attenuate disagreement.

Researchers of children's social development have increasingly utilized approaches employed in pragmatics and sociolinguistics (for example, see Gottman & Parkhurst, 1979; Rubin & Borwick, this volume). At the same time, many researchers of pragmatics in children's language development have become interested in identifying the range of speech acts that children employ at different ages. In assessing peer interchanges, sociolinguists have focused more on the use of speech acts in diverse social contexts and have emphasized the importance of examining role and status factors such as age and sex (see Wilkinson, Clevenger, & Dollaghan, 1981). In general, researchers have been interested in analyzing a given statement for both its stated content and social purpose. A statement such as "I need some help" thus is an account of the speaker's situation and also a social directive that constitutes an indirect request for help from the listener.

The social purpose of a speech act is determined largely on the basis of the nature of the speech event, including status and role considerations. Thus, "I need some help," if said in a situation where the speaker is learning a task from the listener, would represent an indirect request for further instruction. In a situation where one friend is telling another of some personal difficulty, it might reflect an indirect request for emotional support or advice. Knowledge of the appropriate use of speech acts in diverse situations is thus important to effective communication. Continued study of diverse types of speech acts employed by children in conversations, and more specifically in conversations within specific social contexts (for example, with a peer in a particular type of situation) should provide more information about the role of communicative ability in children's peer relationships.

Some of the many research directions in pragmatics and in sociolinguistics are especially relevant, not only for the specification of diverse speech acts such as requests or commands, but also for conversational skills that have been investigated previously with adults and are now being examined with children. One basic conversational skill is that of turn-taking in speaking (Ervin-Tripp, 1977), a skill that even very young children have learned. Beyond simple turn-taking, children soon learn rules for initiating, sustaining, and terminating conversations. Integral to conversations are techniques built on the existence and presumption of shared meaning, including specific contextual meanings and information within the conversation per se. Mishler (1979) has also proposed that it is important to consider the structural boundaries defined by the overall purpose of the conversation, for example when bargaining in trading snacks. Mishler's analyses of 6-year-old children's conversations while bargaining to trade cookies and popcorn revealed that the conversations were structured around not giving the upper hand to one's peers, which was signaled by being the first one to offer a specific trade.

Investigators have also examined the effects of age and situational differences on children's communication skills. Ervin-Tripp (1977) has reviewed evidence indicating that even 5-year-old children can identify the rank and status of a speaker and are sensitive to various social features of a social context (for example, familiarity with the speaker). In addition, children of this age were also found to differentiate a wide range of social directives in their speech, including different types of imperatives, questions, and need statements.

Based on several studies, it appears that elementary school children may use some types of directives far more often than others. Mitchell-Kernan and Kernan (1977) found that when 7- to 12-year-old children were role-playing with puppets, their conversations were dominated by a frequent use of imperative or demand-type social directives. A high frequency of imperatives was especially typical among equal-status peers and from a higher-status speaker to a lower-status listener. Blumenfeld and Kinghorn (1978) studied elementary schoolchildren's persuasive behaviors and found imperatives to be the most common directive or strategy. Wilkinson et al. (1981) found that in classroom groups, children used direct and indirect forms of request, requested information, made statements that mitigated or aggravated a conflict, or used refusals or accounting (rationales) in their peer interactions.

The present research was designed to contribute further to investigations of children's communication and cooperation strategies. The codes employed were derived in part from previous research, most specifically Gottman and Parkhurst (1979) and Ervin-Tripp (1977), and from our own analysis of the structure of the social activity context. The specific aim was to gain information about children's communication strategies in a conflict situation and to examine age and sex differences. Conflict situations have been found to present challenges for children in their peer relationship development (Gottman & Parkhurst, 1979), especially for children with problematic peer relationships (Coie et al., 1982). Comparisons of children with low versus high peer acceptance on the communication skills that are identified to be important can then be examined in future studies.

The present research is derived from two databases, one from a conflict-of-interest session used in a study that focused on the social proceses involved in children becoming acquainted (Oden, Herzberger, Mangione, & Wheeler, 1983), and the other from a study of best friends and friends during the same conflict situation (Wheeler, 1981). For the basic database, we videotaped the dyadic interaction of five-, six-, and seven-year-old same-age, same-sex peers who were newly acquainted from two to three previous ten-minute play sessions conducted on successive days. The conversations of these newly acquainted children were then compared to the second database that consisted of the conversations of classmates who either mutually liked each other or were mutual best friends.

OVERVIEW OF THE STUDY

Same-age and same-sex pairs of unacquainted children at several middle-class schools were selected from a pool of five-, six- and seven-year-old children resulting in nine male and ten female dyads. The children ranged in age from five years and three months to seven years and five months, with a mean dyad age of 76 months (six years and four months). Children who were paired were within six months of each other in age and did not attend the same school.

Conflict-of-Interest Session

Procedure. The study took place in a room equipped with a two-way mirror. Videotaping of the children's interactions took place behind the mirror. Children played with art materials (magic markers, crayons, and so on) that were provided. One-half of the dyads had two ten-minute play sessions on successive days prior to the conflict-of-interest session, and one-half had three sessions. All children played with the same partners each time.

For the conflict session, children were introduced and instructed in the use of a toy (Spin Art) for making designs. In a previous pilot study, this toy had been found to be both enjoyable and to have inherent resource constraints, since only one picture could be made at a time. The Spin Art toy was battery-operated and rotated a square sheet of paper so that circular designs could be made using marker pens.

Coding scheme. To evaluate the children's cooperative and communicative strategies, two coding schemes were developed: social activity roles and communication strategies. Children's conversations were transcribed verbatim from the videotapes. The social activity role coding scheme (see Table 5.1) was developed based on our interpretation of the possible social activity roles that could be enacted to structure this type of conflict-of-interest situation. Children could interact cooperatively and share the resources (the Spin Art toy) by taking turns, or óne child could try to make use of the Spin Art toy exclusively. Dyads could make pictures together and then decide how to divide the pictures, or they could enact various cooperative structures for all or parts of the session. The time limit (ten minutes), however, was also a constraint to be considered, and if turns were taken, children also had to decide when and how to switch turns within the time frame.

A child was judged to be in the *primary role* when observed to have his or her pen marker on the paper, or placed just above the paper, or when seated in a way that made him or her clearly in charge of the activity. In the *secondary role*, a child could still participate in the activity, but in a variety of secondary "jobs": observing, turning the machine off and on, assisting with suggestions, and so on. A child in this role might be helpful, interested, supportive, or mainly trying to gain the primary role. A child might not participate in the secondary role, instead playing with other available art materials while waiting for a turn. A dyad might also enact a *joint role* interaction for all or part of the session. In the joint role structure, the children cooperated by making designs together.

For each child, turns in the primary, secondary, joint, or other (for example, playing another activity alone) role were measured

TABLE 5.1 Coding for Social Activity Roles and Communication Strategies

	Definition or Example
Social activity roles	
primary role	frequency of Spin-Art turns
primary role duration	duration of turns per session
primary turn discrepancy	turn discrepancy between partners
primary duration discrepancy	duration discrepancy between partners
secondary role	types of participation
dispute frequency	frequency of disputes, conflicts
dispute duration	duration of disputes, conflicts
Communication strategies	
speech acts	
requests	"Let's share the pens."
demands	"Stop it!"
agreements	"Okay, fine."
refusals	"No, I won't."
statements	"This is fun."
Social content	
partner monitor/help	"Don't stop it that way."
self plan/monitor	"Now I'm doing blue."
evaluation	"Wow. Pretty."
rationale	"Because I only had one turn."
rule	"You have to share."
joke	"Oh, weirdo!"

by their frequency and duration. Measures of discrepancy for each dyad were obtained by calculating the difference between partners in the number and duration of primary role turns, yielding two discrepancy scores. The number and duration of overt disputes within the overall conflict-of-interest situation were also coded. A dispute was defined as a difference about an issue, viewpoint, procedure, job, or turn that was marked by discordant affect. A dispute, in other words, was a quarrel, argument, or conflictual discussion, as distinct from a simple difference of opinion.

A second set of codes, communication strategy codes, were developed to assess the smallest meaningful units or speech acts at two levels for each speech act at two levels for each speech unit: (1) the type of speech act (such as demand or invitation) and (2) the type of social content (such as evaluation or instruction). Several utterances at a time could be coded on the transcripts after listening to and watching a small segment of the videotapes.

Table 5.1 shows the codes that were found to occur at least one percent of the time and that were reliable. One investigator served as a reliability coder for comparison with the one for the social activity role codes. Two different observers were used for the communication codes, and reliability was based on ten dyads that both coders assessed. Reliability correlates for the codes in Table 5.1 were significant with $p < .05$ as the criterion. The average coefficients were as follows: social activity roles, $r = .76$; speech acts, $r = .82$; social content, $r = .79$.

Social Interaction Processes

The social activity role and communications strategy codes represent the kinds of communication and cooperation skills that appear to be important in this type of social encounter. It was of interest to determine how the dyads generally structured the activity and the kinds of specific communication strategies, including the type and content of the speech acts. Since the dyad was considered as the unit of analysis, a sum score of frequencies for each measure was thus calculated and employed in the statistical analyses. To examine age differences, an average age score was calculated for each dyad and correlated with all measures. Dyads were also divided into younger versus older groups of male and female dyads according to a median split of 77 months, resulting in nine younger dyads (four male and five female), and ten older dyads (five male and five female). A series of sex (2) x age (2) analyses of variance were conducted on all measures.

Primary and secondary roles. Missing data for the social activity role codes for one dyad resulted in eighteen dyads (nine male and nine female). Only one young female dyad used the Spin Art toy by making designs jointly or together for the entire ten minutes, and in another young female dyad, one child used the toy exclusively. These two dyads were also not included in the statistical analyses. All other dyads took turns in the primary role with Spin Art at least once, with an average of 2.03 primary-role turns per child. There were no significant age or sex differences found in the number of primary-role turns. For each child, the average primary-role turn duration was two minutes, with an average of 241.53 seconds (4.03 minutes) in the primary role per session. The following examples from the transcripts illustrates a typical discussion regarding turn-taking.

Child A	Child B
When you gonna be finished?	
	About. . . Okay, now, just this, that, and these two. These two, okay?
(no reply)	
	(about a half minute later) Okay, now turn it off.
That the end?	
	(nods)
Good, now my turn.	

Male dyads were significantly higher than female dyads in their primary role duration (M = 268 seconds for males; M = 211.75 seconds for females, F(1, 13) = 6.48, p<.02). Younger dyads were significantly lower than older dyads in their duration in the primary role (M = 208.75 seconds for younger dyads; M = 264.60 seconds for older dyads, F(1, 13 = 6.28, p<.02). Males and older dyads, in general, may have been more task-oriented or efficient in switching turns than females and younger dyads, or simply more eager to get their turn.

The children's participation in the Spin Art activity while waiting for a primary turn role turn varied considerably. The range of participant activities observed by the coder included watching the partner, participating by turning on and off the Spin Art toy, helping to select colors, handing over markers, making suggestions, and evaluating picture quality. In some cases a child might have been invited to participate in a specific way ("You can turn it on for me"; "Now give me a different color"). In other cases, a child would claim a specific job ("What color do you want? or "I'll turn it on").

With occasional exceptions, children rarely played alone while waiting their turn for the primary role. As indicated earlier, one dyad, a young female dyad, played jointly (marking designs together) for the entire session. Twelve dyads (six male, six female) never played jointly, thus leaving only four dyads who played jointly at least part of the time. The primary and secondary roles were thus the predominant activity roles enacted by the children in this conflict-of-interest situation.

Discrepancies and disputes. Children's references to turn-taking in the primary role indicated their awareness of discrepancies in turns. The following are typical statements from several dyads indicating concern about the ten-minute time limit and perceived discrepancies in the number of turns, or pictures:

> "Yeah, but I won't be able to do anything 'cause you're taking too much time."
>
> "I'm not done, okay?" "You did more than I did."

Discrepancy scores were achieved by calculating the difference in scores between children in each dyad for the number and duration of turns. Eight dyads had no discrepancy in the total number of turns. The average turn discrepancy of the dyads with discrepancies was 1.11. Nine dyads were above the median duration discrepancy of 56 seconds; three dyads had 1.5 minutes or less, and six dyads had discrepancies ranging from three to ten minutes. Nearly equal numbers of male and female, younger and older dyads were found below and above the median. Although the analysis of variance was not conducted due to low cell sizes, an inverse linear relationship between age in months and duration discrepancy was found ($r = .42$, $p<.05$). This finding indicates that greater age was related to less turn discrepancy. Older dyads may have been more efficient in turn-taking, as indicated earlier, or else more cooperative.

Disputes were observed to center frequently on taking a turn with the Spin Art toy. Disputes also occurred over picture style, color, quality, or technique. The following examples from the transcripts illustrate typical disputes observed:

Child A	Child B
I'm gonna do it.	
	I'm gonna do it.
You're not gonna do it!	
	Yeah I am.
It's mine! Martin.	
	This is mine!!

Would you—all right, I quit!
I'm not doin' nothin'.

But you got in. You had a
turn.

You always—I didn't even do
anything!

Yeah you did.

What? Nothin'—I didn't . . . I
didn't do nothin'.

Yeah you did.

No, I didn't. Mart . . . all
right!

All that . . . all that I was
tryin' to do . . .

Well, let me have a turn!

All that I was tryin'—you got
a turn.

I DID NOT!!

Yeah, you did.

Child C

Child D

No, no, no!

Turn it off.
Hey, how'd it do that?

Don't turn it off yet.

I wanna do one now. You're
done.

I'm not done.
Phil, ya gotta let people do
stuff a little more, okay?

I think . . .

Oops. Ooh.

Huh? Now you better turn
that thing off.

No, not that, not yet. Now
I'm gonna put this green on.

Now.

Overall, seven dyads had no disputes. These dyads included
nearly equal numbers of older and younger male and female
dyads. The average number of disputes for those eleven dyads who

did have disputes was 2.36, with an average duration of 75 seconds (one minute and fifteen seconds). These eleven dyads included two younger and three older male dyads, and two younger and four older female dyads. Although the cell sizes limited statistical examination, it appears that more of the older dyads tended to have at least one dispute.

The pattern of the means for the male and female dyads who did have disputes indicated that males engaged in significantly more disputes than females (M = 3.00 versus M = 1.83; t(9) = 1.95, p < .05) and for a significantly greater amount of time (M = 113.2 seconds for male dyads and M = 42.33 seconds for female dyads; t(9) = 2.47, p < .05). It was next of interest to consider whether or not disputes were contributing to more effectiveness in equalizing turns and time with the Spin Art toy.

High- Versus Low-Discrepancy Dyads

To examine the relationship between turn duration and disputes, dyads were divided at the median of the primary role discrepancy distribution (Mdn = 56 seconds). Two dyads, one at each extreme— the one that played with the Spin Art jointly and the one in which the child played exclusively with the toy were not included, resulting in eight dyads of nearly equal numbers of younger and older, male and female dyads in both groups. The low duration discrepancy dyads were found to engage in significantly fewer disputes (M = .75) compared to the high turn discrepancy dyads (M = 2.50; t(14) = 2.97, p < .01). The average dispute duration of the low-discrepancy dyads was also significantly less (M = 19.25 versus 83.25; t(14) = 2.51, p < .05). Duration discrepancy in the primary role turns thus appeared to be very strongly related to dispute interactions between dyads. Apparently, these discrepancies were very evident to the children and appeared to be sources of dissatisfaction. It may also be that the dyads with the longer disputes ran out of the time needed to equalize turns with the Spin Art toy. The Spin Art toy was useful in discriminating those children who had difficulty with the inherent conflicts-of-interest and may be useful in future studies involving cooperation among children.

The speech act and social content codes revealed the communication strategies in this conflict situation. These data will first be described and then related to the social activity role codes in order to evaluate the interface between children's cooperation and communication processes.

Communication Strategy Codes

As indicated previously, the communication codes included two levels of codes for each utterance: the type of speech act and the social content. Data for all nineteen dyads were available for the analyses. Social content codes that occurred less than 1 percent of the time included behavioral threat, affection withdrawal threat, physical contact, compromise, reconcile, and other. Social content codes that could not be reliably assessed according to this criterion included general sociable, dyad planning, partner planning, activity monitoring, self-evaluation, and fantasy. Table 5.2 shows the remaining codes.

Speech acts. Two basic types of speech acts were used 30 percent of the time: requests and demands. Requests were the most frequent form. Responses included agreements and refusals, with agreements somewhat more frequent than refusals. Requests were found to be positively correlated with agreements ($r = .56$, $p < .05$), indicating that the dyads who tended to utilize a greater frequency of requests were also more agreeable. Statements were general utterances, including questions, but not requests, demands, agreements, or refusals. Statements constituted nearly one-half of the utterances (45 percent), and questions made up approximately 8 percent of this total. No other significant correlations were found between the speech act codes. Overall, the form of children's speech acts in this conflict-of-interest situation was rather cooperative and polite, despite dispute episodes. The following example from the transcripts shows this basic pattern:

A young female dyad

Here's the light brown.

No, I don't want to use crayons. Do you mind if I use, uh, these?

Nope, here I decorated it.
Now put the pink back.
I'll use blue.

Now . . . I'll use red . . . if I can get it.

Don't you like the color
pink?

Yeah, but I need red.

An older male dyad

You turn it on here.

I know. I'm just looking.
Do you want any more
markers?

Want this? Here's some.

Okay.

Do you want this color?

I don't know.

Oh, jeepers, look at this!

Oh.

Tell me when you're
done.

Okay.

The pattern of means for speech act codes indicated some interesting age and sex differences. For requests, a significant age
× sex interaction ($F(1, 15) = 7.40$, $p < .02$) was found, indicating
that older females made fewer requests than younger females. The
effect of age seems especially strong for males, also indicated by
a significant positive correlation for males between age (in months)
and number of requests ($r = .76$, $p < .05$). A trend for sex differences was found for demands, with males in generally utilizing
more female demands than females ($F(1, 15) = 4.07$, $p < .06$). At
the same time, however, older females appeared to make greater
use of demands, as indicated by the significant positive correlation between age and demands ($r = .56$, $p < .05$).

Agreement responses indicated a trend for sex and age difference, with more females than males utilizing agreements ($F(1,$
$15) = 4.06$, $p < .06$). Still, for male dyads, agreements increased
as age increased, as indicated by a highly significant correlation
between age and agreements ($r = .92$, $p < .05$). Significantly more
older than younger dyads also used agreements ($F(1, 15) = 5.67$,
$p < .05$). The use of refusals indicated no sex or age differences,
with all groups making use of this form of utterance.

TABLE 5.2 Means and Percentages of Communication Strategies

	M	SD	%
Speech acts			
requests	31.74	12.14	18
demands	21.84	10.93	12
agreements	21.42	9.99	12
refusals	14.42	6.40	9
statements	84.21	23.37	45
Social content			
partner monitor/help	60.37	15.42	33
self plan/monitor	47.58	11.52	27
evaluate	16.00	7.72	9
rationale	15.58	8.82	7
rule	1.47	1.84	1
joke	1.26	2.71	1

NOTE: N = 19 dyads; means are based on the sum of scores per dyad.

While males tended to use more demand communication strategies compared to females, there appeared to be a shift with age toward a less confrontational communication style. Alternately, while females tended to respond with more agreements than males, older females used fewer request communication strategies compared to younger females, and more demand communication strategies. Research findings often have indicated that even young children's social behavior is influenced by sex-role expectations, with males tending toward more aggressive behavior than females (see Fagot & Kronsberg, 1982). It may be that with developmental progression, each sex adjusts or corrects its interactional style toward one that is more effective for a given situation or task.

Social content. Table 5.2 also shows that the most frequent social content codes included *self planning/monitoring* and *partner monitoring*. These social behaviors can be seen in the following examples:

Self Planning/Monitoring
"Yes, I only have three colors. Yep, I do it different. I only have two colors, this one and that one . . ."

"I'm gonna have to set this on, the other side . . . Now, I'm gonna use this—cool."

Partner Monitoring/Helping

 Child A: How does that do that?

 Child B: It's easy, it just turns around and ya' do it. Okay?

 Child A: What—daya, just go like this?

 Child B: Yeah.

Evaluation behaviors occurred less often but seemed to add considerably to the affective tone of these interactions: "Oh, mine's horrible . . . Isn't it ugly?" "Blue and purple?" "Neato." "That turns neat." "Oh, wow." "Now let's see what it looks like." "Aw, cool."

Rationales occurred at a similar level as evaluations and pertained to reasons for color selections and to reasons for starting out, retaining, or reclaiming a primary role turn: ". . . cause I didn't want that color in it." "Hurry! We're not gonna have much time to play this game." "There, this is your last color, or it starts spinning back." "No, wait. I gotta do these." "Yeah, but I want to use red."

Rules, joking, and dyad monitoring utterances occurred only 1 percent of the time, although this percentage may greatly underestimate their importance in the interaction. Upon reading the transcripts, it appeared that the use of rule citing ("We have to share . . . You can't always have your way") was used as a strategy for getting one child to wait for his next turn on the Spin Art toy. Analyses that are based on frequencies and proportions may underestimate the role of low-frequency behaviors in social interaction. Analyses of variance computed for each of the social content codes revealed no significant main effects or interactions for age and sex.

Social Activity Roles and Communication Strategies

Some interesting relationships were found between the two sets of codes that examined children's cooperative and communicative strategies. Demand strategies were positively correlated with primary role turns ($r = .50$, $p < .05$) and refusal strategies were positively correlated with primary turn discrepancies ($r = .52$, $p < .05$) and the number and duration of disputes ($r = .68$ and $.59$, $p < .05$). An assertive pattern of speech acts was thus found to be related to gaining primary role turns. Refusal speech acts ("No!" "Not yet") appeared to signal an unyielding stance in situations of maintaining the primary role and in conversations marked by disputes. Furthermore, monitoring and helping were positively correlated with several social activity role variables: primary role turn frequency and duration (r = .58 and .44, p < .05), secondary role duration (r = .67, p < .05), and number of disputes (r = .44, p < .05), but not duration of disputes. From the transcripts, it appeared that children in the secondary role were often helpful to the point of being an interference, leading to disputes over which colors to use and ways to utilize the pen markers. At times, one child appeared to be hovering over another child, waiting anxiously for his or her turn. Evaluative strategies, however, were found to be inversely related to the number of disputes (r = .41, p < .05), indicating that evaluations were received positively and may have contributed to a more cooperative interaction.

Type of Peer Relationship

It was of interest to consider that the cooperative and communication strategies used by the newly acquainted dyads might differ from those of children in peer relationships that were more established and known to be reciprocally positive. We thus compared the six and seven-year-olds in our basic sample of recently acquainted dyads with first-grade classmates who participated in another study (Wheeler, 1981). In this latter study, the children were also provided with the same Spin Art toy for a ten-minute period and were videotaped. The dyads were selected on the basis of sociometric analyses conducted several weeks prior to the Spin Art play session. The dyads selected for this study were paired on the basis of reciprocal liking and friendship. All children in this sample were well liked by their peers and paired with a classmate they liked or who was a mutual best friend. For the present analyses, the newly acquainted children (n = 12) from the present study were compared to the "friendly" classmates from the Wheeler study (n = 24). Differences between these groups were found for social activity role and communication strategy codes.

Partners within the classmate dyads switched turns more frequently with the Spin Art toy (M = 4.33) than did the newly-acquainted dyads (M = 2.04; t(34) = 4.67, p < .001). However, there were no significant differences in primary role turn or duration discrepancies, or in the number of disputes. These findings are interesting, in that we might expect that friendly peers would be more equitable. Still, the disputes were of less duration for classmates than for acquaintances (M = 10.38 seconds versus M =

45.08; t(34) = 3.43, p < .01). In general, it appears that children who have a more developed relationship may be more likely to accept more discrepancies in resource distribution.

Due to coding differences and reliability limitations, some communication and strategy codes were not included for comparison. Greater use of direct speech acts such as demands was found for classmates than for newly acquainted dyads (21 percent versus 12 percent). Classmates, however, used somewhat fewer refusals (8 percent) than newly acquainted dyads (5 percent), who also used a somewhat higher frequency of agreement responses (12 percent versus 9 percent). Overall, social content codes indicated that classmate dyads monitored each other's activities more frequently than the newly acquainted dyads, (7 percent versus 2 percent). This latter finding may be indicative of the greater length of disputes for the newly acquainted dyads.

Wheeler's (1981) study of six-year-old best friends and classmates revealed a pattern of relationships between primary role discrepancies, disputes, and communication strategies that was highly similar to the database of newly acquainted dyads. In the Wheeler study, however, there were no significant correlations for evaluation strategies. Fantasy play, however, was inversely related to the number of disputes, which suggests that a more playful interchange may help to create fun and a casual atmosphere, thereby lessening the likelihood of disputes. This type of strategy may be more prevalent among friends and well-acquainted dyads, a finding observed in previous research (for example, Gottman & Parkhurst, 1979).

In the Wheeler study, another pattern of significant correlations indicated that more primary role turns, less primary role duration, and the number and duration of disputes were related to providing rationales or reasons as to why the speaker should gain or maintain the Spin Art primary role. Although rationales were effective strategies for gaining greater use of the Spin Art toy, this was not generally well received by partners. Providing a rationale may have been viewed as simply another strategy for gaining greater use of the Spin Art toy.

DISCUSSION

Developmental Implications

A major question that emerged from this research with regard to children's cooperation concerns the basis of the dyadic discrepancies in the number and duration of primary role turns with the Spin Art toy. It may be argued that highly turn-discrepant dyads were likely to have one or both partners who were developmentally less competent compared with their peers. In this study, age was found to be inversely correlated with primary role duration discrepancies. However, magnitude of the correlation, although significant, was not sufficient to rule out additional factors. Furthermore, there were nearly equal numbers of younger and older dyads in both the low- and high-discrepancy dyads that were differentiated.

Wheeler's (1981) postexperimental interview data indicated that while many children did understand that the social interaction process could be optimized by strict reciprocity or equity in taking turns for Spin-Art use, other children indicated no response or lacked such understanding. Berndt (1982) reviewed data by Piaget (1932/1965) and others and proposed that even young children who do understand and accept the principle of equality of turns may try to rationalize more turns on some other basis — for example, their picture did not turn out right, so they should have an extra turn. Thus, some children may not actually employ the equity principle.

Upon reading the children's conversations in the transcripts from the present research, and based on the data analyses, it appeared that the children often understood that there was an inequity but tried to maneuver around the issue. Some children, however, appeared to think that the basis was negotiable. Many children were found to be quite assertive in their negotiations. Additional evidence reviewed by Berndt (1982) indicates that by seven years of age, children do no accept the strict equity principle (strict equality of turns, for example) as the sole principle in resource distribution. Children of this age appear to be capable of considering principles that take into account the needs of others and individual differences. Furthermore, research with children of similar age by Forbes and Lubin (this volume) indicates that children's social behavioral strategies are related, at least in part, to an active consideration of their social knowledge relevant to a situation.

In negotiating, some children's communication strategies may be inappropriately competitive or aggressive due to a developmental lag, poor social cognitive or information processing, or previous experiences. In general, younger children under approximately the age of nine or ten years have been found to be more competitive in peer interactions, even when paired with close friends (see Berndt, 1982), although mutuality has been found with some tasks among friends (Newcomb, Brady, & Hartup, 1979; Newcomb & Brady, 1982), and a high tolerance of discrepancies in strict reciprocity was found in the present research among classmate dyads.

Dodge and Frame (1982) have presented evidence that aggressive children may lack realistic interpretations of their peers' behavior toward them; that is, they may tend to view their peers' behavior as hostile when it is not. The basis of such social information processing difficulties may originate with the family through modeling of similar interpretations or experiences that involve high degrees of aggression (Herzberger & Tennen, 1982). Some family experiences also emphasize the value of competition. One source of difficulty for those children with less effective dyadic interactions may be a lack of skill in selecting strategies that are appropriate to a given situation in terms of peer norms or expectations, perhaps due to a lack of experience in peer relationships, or to excessively negative interactions (see Oden, 1980).

The role of disputes in conflictual situations needs further examination. In the present research, it may be that disputes actually added to discrepancy problems by diminishing the time to be divided. Participation in disputes may have been a stalling tactic to maintain use of the Spin Art toy, or a mechanism for resolving turn discrepancies and disagreements about task procedures. Disputes may have also been an indication of incompatibility or dislike between dyadic partners. A demand-refusal communication strategy pattern was found to be related to more disputes and less equality or reciprocity in use of the Spin Art toy. Some children may have also learned a limited and/or aggressive interactional style that achieves the sought-after goal, but perhaps without the awareness that it is with the likely result of aggravation or alienation of peers. These children may lack a range of alternative communication strategies and/or awareness of all the possible consequences, or they may be operating from a different set of values. Some children may, for example, be developmentally younger, or they may not have learned to value fairness as much, although again they may not be aware of the consequences. Others may value satisfying their individual needs or goals more than positive peer relationships and lasting friendships.

Future Research Directions

Future research should be aimed at gaining more information on the relationship between children's communication strategies in conflictual stituations and their awareness of the consequences for various specific strategies. Children should also be observed with several peer partners in order to separate the effects that may be due to unique pairings and to discern some regularity in dyadic interactional styles in such situations. Studies should also focus on comparisons of interactional styles for individual children as they interact in diverse types of relationships—for example, with a best friend, classmate, or acquaintance.

The observations and interviews employed in future studies should also be conducted with children of various levels of peer status in order to identify the kinds of communication strategies and social cognitive processes that may be lacking in children of low peer status. Rubin and Borwick (this volume), for example, found that four- and five-year old children, who were identified as socially isolated and subsequently observed in various dyadic peer pairings, utilized different types of request strategies compared to more sociable peers. In the present research, although no sociometric data were collected for the basic database, and while only well-liked children were used in the Wheeler (1981) study, we might expect that children who lacked negotiation skills or who were not cooperative in resource distribution would be less accepted by peers.

It is interesting to consider that the frequency or appropriateness of communication and cooperation skills of accepted versus unaccepted peers may not be extremely discrepant, but only to a more moderate extent as indicated by Rubin and Borwick's data. Indeed, the present research found only a few children to be highly uncooperative. Instead, in children's peer contexts, such as the classroom, there may be a threshold of acceptance for behavior that deviates from the norm, or from average levels or types of communication and cooperation. It may be that social skill training is effective when it results in helping the less accepted child to learn how to stay within that threshold of acceptance and also provides social participation experiences so that changes in levels or types of behaviors may be noted by the more accepted peers (see Oden, in preparation). Recent research by Singleton (1981) indicates that peers may be receptive to perceived changes in a child's behavior. Future studies that examine varying levels of children's peer status in order to compare their cooperative and communication strategies may yield additional information for designing scripts or methods for coaching in social skills (for example, Oden & Asher, 1977; Ladd, 1981) and in social problem solving (see Urbain & Kendell, 1980).

In conclusion, the present study has demonstrated that communication strategies are meaningfully related to children's cooperation in situations with inherent conflicts of interest, and that children vary considerably in their ability to cooperate effectively in such situations.

REFERENCES

Applegate, J. L., Burke, J. A., Burleson, B. R., Delia, J. G., & Kline, S. L. (in press). Reflection-enhancing parental communication. In I. E. Sigel (Ed.), *Parental belief systems: The psychological consequences for children.* Hillsdale, NJ: Lawrence Erlbaum.

Berndt, T. J. (1982). Fairness and friendship. In K. H. Rubin and H. S. Ross (Eds.), *Peer relationships and social skills in childhood.* New York: Springer-Verlag.

Blumenfield, P. C., & Kinghorn, S. H. (1978). *Situational influences on children's persuasive behaviors.* Paper presented at the annual meeting of the American Educational Research Association, Toronto.

Coie, J. D., Dodge, K. A., & Coppotelli, H. (1982). Dimensions and types of social status: A cross-age perspective. *Development Psychology, 18,* 557-570.

Dodge, K. A., & Frame, C. L. (1982). Social cognitive biases and deficits in aggressive boys. *Child Development, 53,* 620-635.

Ervin-Tripp, S. (1977). Wait for me, rollerskate! In S. Ervin-Tripp & C. Mitchell-Kernan (Eds.), *Child discourse.* New York: Academic Press.

Fagot, B I., & Kronsberg, S. J. (1982). Sex differences: Biological and social factors influencing the behavior of young boys and girls. In S. G. Moore & C. K. Cooper (Eds.), *The young child: Reviews of research: Vol. 3.* Washington, DC: National Association for Education of Young Children.

Gottman, J. M. (1977). Toward a definition of social isolation in children. *Child Development, 48,* 513-517.

Gottman, J., Gonso, J., & Rasmussen, B. (1975). Social interaction, social competence and friendship in children. *Child Development, 46,* 709-718.

Gottman, J. M., & Parkhurst, T. J. (1979). A developmental theory of friendship and acquaintanceship processes. In W. A. Collins (Ed.), *Minnesota symposium on child psychology: Vol. 13.* Hillsdale, NJ: Lawrence Erlbaum.

Hartup, W. W. (1978). Children and their friends. In H. McGurk (Ed.), *Issues in childhood social development.* London: Methuen.

Herzberger, S. D. & Tennen, H. (1982). *The social definition of abuse.* Paper presented at the American Psychological Association, Washington, DC.

Krantz, M. (1982). Sociometric awareness, social participation and perceived popularity in preschool children. *Child Development, 53,* 376-379.

Ladd, G. W. (1981). Effectiveness of a social learning method for enhancing children's social interaction and peer acceptance. *Child Development, 52,* 171-178.

Ladd, G. W., & Oden, S. (1979). The relationship between peer acceptance and children's ideas about helpfulness. *Child Development, 50,* 402-408.

Mishler, E. G. (1979). "Would you trade cookies with the popcorn?" The talk of trades among six year olds. In O. K. Garnica & M. L. King (Eds.), *Language, Children and Society.* New York: Pergamon Press.

Mitchell-Kernan, C., & Kernan, K. T. (1977). Pragmatics of directive choice among children. In S. Ervin-Tripp & C. Mitchell-Kernan (Eds.), *Child discourse.* New York: Academic Press.

Newcomb, A. F., & Brady, J. E. (1982). Mutuality in boys' friendship relations. *Child Development, 53,* 392-395.

Newcomb, A. F., Brady, J. E., & Hartup, W. W. (1979). Friendship and incentive condition as determinants of children's task-oriented social behavior. *Child Development, 50,* 878-881.

Oden, S., & Asher, S. R. (1977). Coaching children in social skills for friendship making. *Child Development, 48,* 495-506.

Oden, S. (1980). The child's social isolation: Origins, prevention, intervention. In G. Cartledge & J. Milburn (Eds.), *Teaching social skills to children: Innovative approaches.* New York: Pergamon Press.

Oden, S. (in preparation). *An alternative perspective on children's peer status and social skills.*

Oden, S., Herzberger, S. D., Mangione, P. L., & Wheeler, V. A. (1983). Children's peer relationships: An examination of social processes. In J. C. Masters & K. Yarkin-Levin (Eds.), *Boundary areas in social and developmental psychology.* New York: Academic Press.

Piaget, J. (1965). *The moral judgment of the child.* New York: Free Press. (Original work published 1932)

Putallaz, M., & Gottman, J. (1981). Social skills and group acceptance. In S. R. Asher & J. M. Gottman (Eds.), *The development of children's friendships.* New York: Cambridge University Press.

Rubin, K. H. (1972). Relationship between egocentric communication and popularity among peers. *Developmental Psychology, 7,* 364.

Rubin, K. H., & Everett, B. (1982). Social perspective-taking in young children. In S. G. Moore & C. R. Cooper (Eds.), *The young child: Reviews of research: Vol. 3.* Washington, DC: National Association for the Education of Young Children.

Singleton, L. C. (1981). *The influence of reputation on children's perception of behavioral change.* Paper presented at the annual meeting of the American Psychological Association, Los Angeles.

Tesch, S. A., & Oden, S. (1981). *Referential communication skill and peer status in third and sixth grade children.* A paper presented at the annual meeting of the American Psychological Association, Los Angeles.

Urbain, E. S., & Kendall, P. R. (1980). Review of social-cognitive problem-solving interventions with children. *Psychological Bulletin, 88,* 109-143.

Wheeler, V. A. (1981). *Reciprocity between first-grade and nonfriend classmates in a conflict of interest situation.* Unpublished doctoral dissertation, University of Rochester, Rochester, NY.

Wilkinson, L. C., Clevenger, C., & Dollaghan, C. (1981). Communication in small instructional groups, a sociolinguistic approach. In W. P. Dickinson (Ed.), *Children's oral communication skills.* New York: Academic Press.

6

Communicative Skills and Sociability

Kenneth H. Rubin and Diane Borwick

A number of developmental psychologists have postulated that peers play a significant causal role in the development of social competence and social cognition. Piaget (1926, 1932), for example, indicated in his early writings that peers influence social cognitive growth in the middle years of childhood and beyond. During the preoperational period, children were believed more likely to interact with adults than with peers. The child-adult relationship was postulated to be one of unilateral respect: The adult provides the child with social norms, including concepts of right and wrong, and the child views these rules as absolute and unchangeable. This particular social climate does little to alter the purported egocentric nature of thought in early childhood.

Authors' Note: *This project was funded by a grant from the Ontario Ministry of Community and Social Services. Thanks go to Judy Mickle and Anne Emptage for their help in collecting, coding, and analyzing the data. We are also grateful to the preschool and kindergarten children and teachers who kindly consented to participate in this project.*

In contrast, the onset of peer interaction produces a new relationship of mutual respect, with opportunities for conflict and negotiation that result from different social perspectives. Such opportunities were viewed as central to the waning of egocentrism and the onset of mature communicative and social cognitive skills. Although Piaget believed that the role peers played did not have significant impact until the middle years of childhood, contemporary researchers have shown that peer conflict, negotiation, persuasion, and role-play do promote developmental change in the preschool and kindergarten years (Burns & Brainerd, 1979; Iannotti, 1978; Rubin & Pepler, 1980). Moreover, not only are developmental changes in social knowledge and competence produced, but also growth in the impersonal knowledge domain (Botvin & Murray, 1975). Thus it appears that Piaget's ideas concerning the significance of peer interaction for social and cognitive development are apropos for the early as well as the middle years of childhood.

Learning theorists also have noted that peer interaction plays an important role in child development. Unlike Piagetian theory, the learning theorist perspective does not provide a unified, coherent statement concerning the effects of peer interaction on specific areas of social development. Nevertheless, researchers working within the learning theory framework have demonstrated experimentally that children can and do affect each other's social behavior. In particular, there have been numerous studies, both experimental and observational, in which researchers have described the many ways in which peers reinforce each other's behavior both directly and indirectly (through modeling).

Included among those types of social behavior that have been affected by direct reinforcement and punishment on the part of peers are aggression (Patterson, Littman, & Bricker, 1967), prosocial activities (Charlesworth & Hartup, 1967), and sex-typed activities (Lamb & Roopnarine, 1979). Social learning studies concerning the effects of observing peer models on the production of aggressive, prosocial, and sex-typed activities have also demonstrated the indirect impact of peers on social development (see Hartup, 1983; Shaffer, 1979, for reviews).

Given the Piagetian and learning theory perspectives, it becomes quite clear that peers play an important role in promoting social development. However, neither perspective provides much information concerning the effects of the lack of peer experiences on social development. In recent years, it has been proposed that children who do not interact with their peers in early and middle childhood are "at risk" for socially related problems (Hartup, 1983; Rubin, 1982a). Initial support for this contention stems from the primate work of Suomi and Harlow (1972), who found that rhesus monkeys raised by their mothers without exposure to their age-mates produced avoidant and aggressive behavior when placed in peer group play situations. Moreover, such abnormal displays of wariness and aggression were long-standing.

Rubin (1982a, b) has recently found that the observed frequency of certain forms of nonsocial play is negatively related to indices of social and social cognitive competence. For example, children who engage in high frequencies of solitary-functional (sensorimotor) and solitary-dramatic activities are less popular, have lower mental ages, and are viewed by their teachers as less socially competent than their peers who engage in low frequencies of such behavior. Moreover, children who have been targeted observationally as "isolates" evidence less cognitively mature forms of play and receive fewer social overtures from their peers than their more sociable counterparts. For example, isolate children display less dramatic play than their sociable age-mates, and when they do exhibit such play, it is more likely to be carried out in the absence of play partners. Given the purported causal significance of sociodramatic activities for the promotion of social, perspective-taking, and problem-solving skills (Rubin, Fein, & Vandenberg, 1983), and given that solitary-dramatic play is predictive of social maladaptation, the behavior of isolate children may mark them as being "at risk" for developmental problems.

Interestingly, Rubin (1982a) has found that isolate children are more likely to suggest that adults intervene on their behalf when they are asked how to solve object acquisition-related social problems. These social problem-solving strategies are suggestive of isolate children's greater dependency on adults. Moreover, socially withdrawn young children are not often approached by their peers (Rubin, 1982a). Hence, the receipt of social initiatives may be taken as an index of peer popularity. Given that peer rejection and popularity have been found to predict later maladjustment (Cowen, Pederson, Babigian, Izzo, & Trost, 1973; Roff, Sells, & Golden, 1972), there is further reason to be concerned about children who do not interact with peers.

In summary, this brief literature review makes it clear that good reason exists to believe that peer interaction plays a significant role in social, cognitive, and social cognitive development. From the limited extant data concerning social withdrawal, it also appears that children who interact with their peers significantly less often than is the norm for their age group may be at risk for later maladjustment. Admittedly, little is known about the psychological concomitants and consequences of social withdrawal in early childhood, but the available data point in the direction of a negative prognosis.

At this point, having suggested that social withdrawal in early childhood may have negative consequences, it may be appropriate to ask why there is so little clinical or developmental data concerning the correlates and consequences of early peer isolation. There are at least two reasons. First, some researchers have not found socially withdrawn children to differ from their more sociable peers on indices of peer popularity and social cognitive development (see Asher & Hymel, 1982, for relevant reviews). Consequently, it is often assumed that a problem simply does not exist for these children. However, we believe that the earlier research concerning social withdrawl is fraught with conceptual problems. For example, researchers typically have not used a stringent, extreme-groups procedure to identify isolate children. In some cases, young children who interact with their peers during "only" 50 percent of the observation period have been targeted as isolates (for example, Keller & Carlson, 1974). In other cases, children whose production of social interactive behavior falls within the bottom third or half of the class have been identified as isolates. These targeting procedures simply do not take into account behavioral, age-related norms concerning the production of nonsocial behavior in childhood. As a result, the targeted children may be quite normal. Thus, the nonsignificant differences reported in the literature may be a product of nonstringent targeting procedures. In the one study in which an extreme-groups targeting methodology was employed to identify isolate children, significant differences in social and cognitive development between children varying with regard to sociability were found (Rubin, 1982a).

A second reason for the relative lack of attention to nonsocial activity in early childhood may emanate from an overriding concern with the high profile forms of social maladjustment. Researchers interested in social behavioral problems have tended to focus their attention on aggressive and/or hyperactive children who present caretakers with immediate home and classroom difficulties (see Campbell & Cluss, 1982). Isolate children may be somewhat "invisible" to their teachers. They do not act aversively or disruptively in their classes. As a result, teachers and peers tend not to rate isolate children negatively on indices of social adjustment (Asher & Hymel, 1982). Parents, too, simply may attribute their children's nonsocial behavior to a personality disposition of shyness. However, given the literature reviewed above, further study of social withdrawal in early childhood is clearly merited. In this chapter we hope to describe additional data concerning the concurrent correlates of social isolation in early childhood.

Two questions guide our present concerns: First, can social skill differences between socially withdrawn children and their more sociable age-mates be identified in early childhood? Second, are these differences indicative of social maladjustment and deficit? If social skill differences can be discovered in early childhood, then ameliorative intervention efforts focused on specifically identified deficits can be implemented prior to the elementary school years. Successful early intervention may prevent the potential negative consequences of isolate behavior during the preschool and kindergarten years.

COMMUNICATIVE SKILLS OF CHILDREN

One social behavior that has been linked theoretically to peer interaction is communication. Piaget (1926) noted that communicative behavior during the preoperational years reflects a primarily egocentric mode of thought. He speculated that the incidence of egocentric communication would decline with increasing peer interaction, conflict, and negotiation experiences. Interestingly, Rubin (1982a) has reported that isolate preschoolers talk significantly more often to themselves or to inanimate or non-present others during dyadic free play than do their more sociable counterparts. Although we now know that socially withdrawn children produce fewer socially directed utterances, these data tell us little about the qualitative nature of those socially directed utterances produced by children who vary with regard to sociability.

In this chapter we will examine one particular form of socially directed communication—the request. As we see it, any communicative attempt to get another to do what you want him or her to do may be considered an attempt to solve a social problem (Delia, Kline, & Burleson, 1979; Forbes & Lubin, this volume; Krasnor & Rubin, 1981; Levin & Rubin, 1983). It is a social problem that places a high demand on the listener for a response. Given that social problem-solving skills have been viewed as indices of social competence, we thought that an in-depth analysis of the phenomenon would provide us with important descriptive data concerning the correlates of sociability in early childhood. The questions that we attempted to answer in this study included the following: (1) Do isolate children produce fewer requests directed at their dyadic free play partners than their more sociable agemates? If isolate children are indeed less socially oriented, one might expect that they would issue fewer requests than more sociable children; (2) Do children who vary with regard to sociability produce qualitatively different requests? For example, isolate children may be less assertive than their more sociable counterparts, and consequently less likely to emit imperatives than question directives; (3) Are socially withdrawn children as likely to obtain compliance as their more sociable peers? (4) What is the sequence of children's behavior following noncompliance with a request? Are socially withdrawn children more likely to give up or to repeatedly use the original request strategies?

One index of social competence is the flexibility with which children attempt to solve their social problems (Krasnor & Rubin, 1981; Levin & Rubin, 1983; Spivack & Shure, 1974). Following failure, the production of request forms that differ from the original strategies is one such example of flexibility. Another example is the ability to know when an unsuccessful request is doomed to failure and thereafter change the subject of conversation or follow the play partner's change of subject.

In summary, through an analysis of children's requests we hoped to discover whether isolate preschoolers and kindergartners were less socially competent than their more sociable peers. Indices of social competence included the ability to produce a variety of initial request forms (for example, imperatives, question directives, declaratives, and permission requests), to gain compliance, and to produce flexible rerequests following noncompliance.

THE TARGETED SAMPLE

An extended description of the targeted sample appears in Rubin (1982a). The original sample comprised 123 preschoolers (M age = 57.76 months) and 111 kindergartners (M age = 65.05

months) who lived in Southwestern Ontario. In all, 71 of the preschoolers and 56 of the kindergartners were females.

The children were observed during free play following procedures outlined extensively in Rubin (1982b). Briefly, each preschooler was observed for six ten-second intervals each day for 30 days; the kindergartners were observed likewise on 25 separate days. Their behavior was coded on a checklist that included the social participation categories originally described by Parten (1932) and revised by Rubin (1982b); that is, solitary, parallel, and group activities. Other relevant targeting categories included unoccupied behavior, onlooker behavior, and active conversations with peers and teachers.

Isolates were those children whose nonsocial behavior (unoccupied + onlooker + solitary behavior) was one standard deviation above their entire age-group mean and 10 percent above their class means for nonsocial behavior. Moreover, isolates produced social behavior (group play + conversations) that was one standard deviation below the entire age-group mean and 10 percent below their class means. In all, 17 preschoolers (six male, eleven female) and 17 kindergartners (six male, eleven female) were identified as isolates.

Sociable children were those whose social behavior was one standard deviation above the entire age-group mean and 10 percent above their class means. Moreover, sociable children produced nonsociable behavior that was one standard deviation below the entire age-group mean and 10 percent below their class means. A total of 17 preschoolers (nine male, eight female) and 18 kindergartners (eleven male, seven female) were identified as sociable children. All remaining children (that is, approximately 70 percent of the sample) were targeted as neutral.

From these targeted groups, five isolate and five sociable preschool children, and a like number of kindergarten children, were randomly selected. The children were paired with same-age, same-sex, nonfriend, "neutral" classmates. Friendship status was ascertained by referring to the children's responses to a sociometric battery. Only pairs of mutually rated "like a little" (as opposed to "like a lot" and "don't like") children formed the dyads. Detailed descriptions of the sociometric procedures appear in Rubin (1982a, 1982b).

The dyads met in a small playroom set up in a laboratory trailer for fifteen minutes on two occasions. The sessions were separated by at least one week's time. In the room were commercial materials designed to elicit both sociodramatic and constructive play (for example, a Fisher-Price Sesame Street clubhouse; plus blocks and dolls).

The children were brought to the room by an adult experimenter and informed that they could play as they wished for a period of fifteen minutes. The experimenter left to "do some work," which in actuality involved recording the entire play session on videotape through a two-way mirror.

Each videotape was transcribed by the researcher who recorded the children's activities. The coding procedure is described at length below. The unit of verbal coding was the utterance or "that unit of speech separated from the following unit by a logical break in speech, a change in topic, and/or a one-second pause" (Rubin, 1979).

The utterances in each transcript were searched in an effort to identify the following types of requests: (a) *direct requests* uttered in the imperative form (for example, "Come here"; "Gimme dat doll"); and (b) *indirect requests*, including question imperatives ("Can you get me the block?"), declaratives ("You hafta bring me some cars"); and inferred requests ("That Cookie Monster over there should be on the slide"). In all cases, requests were taken as implying that the speaker wanted the listener to do something. Only well-informed requests were considered. Consequently, requests muttered to the self or addressed to inanimate objects were excluded, as in Garvey (1975).

Once all direct and indirect requests were coded, their consequences were classified as follows: (a) success—the partner performs the requested action; (b) query—the partner asks for a clarification of the speaker's request ("Huh?"; "Wha'?"); (c) acknowledgment—the partner acknowledges the request but does not perform the requested action; (d) refusal—the partner refuses to comply with the request; (e) no response—the partner does not perform the requested action and makes no indication that he or she has heard the request; (f) total failures—categories (c), (d), and (e).

In an effort to consider how children varying with regard to sociability responded to failure, the coding procedure drawn from Levin and Rubin (1983) was followed. Briefly, only the first rerequest strategy following failure was considered in our coding scheme. Seven categories of rerequest strategies were coded. A rigid strategy consisted of either a verbatim repetition of the original request, or of a minor modification. A modified strategy involved either a statement of the original request with an added explanation, or a switch from indirect to direct speech forms or vice versa. Other follow-up strategies included requests for clarification of the partner's response ("Why?"; "How come?"); self solutions involving the subject's own performance of the requested action; no further response; a change of subject, through which the original requester alters the topic of conversation without reference to the earlier request; and following the partner's change of subject.

Finally, the speaker's intention was coded from a viewing of the videotapes and a reading of the transcripts. Request intentions included the following: elicit action—requests that the partner do something; stop action—requests that the partner stop doing or refrain from carrying out an action; acquire objects—requests for objects in the partner's possession; gain attention—requests for the partner to attend; joint action—requests for the partner to join the speaker in some activity.

Reliability was computed by having two independent observers code the videotaped and transcribed data. Eight full transcripts were coded (that is, two hours of data). Percentages of agreement for identifying requests, coding requests as direct or indirect, coding outcome, and classifying follow-up to failure were 79 percent, 90 percent, 88 percent, and 85 percent respectively.

RESULTS

Dependent t-tests of session differences were computed for all variables. For the most part, there were few meaningful session differences concerning the types of requests produced, the consequences of the requests, the strategies following failure, and the speaker's intention when producing given types of requests. Consequently, all data were pooled for subsequent analyses.

The data in Tables 6.1, 6.2, and 6.3 represent per session averages for each variable of interest. As a first step, the average numbers of utterances and direct and indirect requests were computed. Second, proportions were calculated for each request consequence only for those children who employed particular request forms. For example, eight isolate children used direct requests (Table 6.1). Given those eight isolate children, .54 (out of 1.00) of their imperatives were successful.

TABLE 6.1 Means for Direct Request Data Averaged Across Sessions

	Isolate	Neutral	Sociable	Neu-Iso	Neu-Soc
	(n = 10)	(n = 20)	(n = 10)	(n = 10)	(n = 10)
Total utterances	91.70	133.27	150.33	104.61	161.93
Direct request	4.91	9.10	9.38	6.80	11.40
	(n = 8)	(n = 20)	(n = 9)	(n = 10)	(n = 10)
No wait	.01	.04	.02	.07	.01
Success	.54	.57	.56	.69	.45
Query	.01	.01	.01	.00	.01
Acknowledgment	.01	.01	.03	.01	.38
Refuse	.02	.08	.11	.03	.13
No response	.32	.29	.26	.20	.38
Fail*	.35	.38	.40	.24	.52
	(n = 5)	(n = 15)	(n = 9)	(n = 5)	(n = 10)
Rigid	.09	.14	.11	.18	.12
Modify	.50	.34	.37	.17	.42
Clarify	.00	.00	.02	.00	.00
Self solve	.00	.04	.03	.05	.03
No further response	.31	.32	.33	.45	.26
Change subject	.10	.07	.03	.10	.06
Follow partner	.00	.08	.12	.05	.11
	(n = 8)	(n = 20)	(n = 9)	(n = 10)	(n = 10)
Elicit action	.22	.42	.41	.42	.42
Stop action	.25	.29	.26	.31	.28
Acquire object	.01	.03	.02	.06	.01
Gain attention	.52	.26	.30	.22	.30
Joint action	.00	.00	.01	.00	.00

*Fail = acknowledge + refuse + no response.

Third, proportions were computed for each type of follow-up strategy employed given a request failure. For example, five isolate children experienced failure following the issuance of a direct request (table 6.1). An average of .50 (out of 1.00) of their follow-up strategies were modifications of some sort.

Finally, proportions of the total numbers of direct and/or indirect requests that had given intentions were computed. For example, an average of .25 (out of 1.00) direct requests were "stop action" for the isolate children (table 6.1).

The data in Tables 6.1, 6.2, and 6.3 are presented for isolate, neutral, and sociable children, as well as for those neutral children who were paired with isolates and with sociable children separately.

TABLE 6.2 Means for Indirect Request Data Averaged Across Sessions

	Isolate	*Neutral*	*Sociable*	*Neu-Iso*	*Neu-Soc*
	(n = 10)	(n = 20)	(n = 10)	(n = 10)	(n = 10)
Indirect request	9.90	15.95	16.85	11.88	20.01
	(n = 9)	(n = 20)	(n = 10)	(n = 10)	(n = 10)
No wait	.00	.02	.01	.04	.01
Success	.54	.56	.59	.61	.51
Query	.02	.03	.03	.03	.02
Acknowledgment	.06	.04	.06	.02	.05
Refuse	.05	.10	.14	.07	.14
No response	.28	.25	.18	.23	.27
Fail*	.40	.39	.38	.32	.46
	(n = 8)	(n = 17)	(n = 10)	(n = 10)	(n = 10)
Rigid	.14	.07	.11	.07	.06
Modify	.21	.47	.27	.51	.45
Clarify	.02	.00	.04	.00	.00
Self solve	.02	.01	.03	.01	.00
No further response	.43	.27	.41	.23	.30
Change subject	.10	.07	.07	.08	.06
Follow partner	.03	.11	.08	.10	.12
	(n = 9)	(n = 20)	(n = 10)	(n = 10)	(n = 10)
Elicit action	.26	.28	.22	.31	.26
Stop action	.07	.08	.12	.09	.07
Acquire object	.06	.12	.12	.10	.05
Gain attention	.51	.36	.37	.30	.42
Joint action	.12	.16	.17	.15	.17

*Fail = acknowledge + refuse + no response.

Three sets of t-tests were computed. First, the requests of isolate children and their non-isolate, neutral counterparts were compared. Second, the requests of sociable children and their neutral play partners were compared. Finally, the communicative data for neutral children paired with isolate versus sociable children were analyzed. Taken together, these analyses portray not only the production of requests, but also the reception of requests by the speakers' play partners.

Direct and Indirect Requests

Series of t-tests were computed for the total number of direct and indirect requests produced by the children. The isolate

TABLE 6.3   Means for All Request Data Averaged Across Sessions

	Isolate	Neutral	Sociable	Neu-Iso	Neu-Soc
	(n = 10)	(n = 20)	(n = 10)	(n = 10)	(n = 10)
Requests	14.81	25.05	26.23	18.68	31.41
	(n = 8)	(n = 20)	(n = 9)	(n = 10)	(n = 10)
No wait	.01	.03	.02	.05	.01
Success	.54	.56	.57	.65	.48
Query	.02	.02	.02	.02	.02
Acknowledgment	.03	.02	.05	.01	.03
Refuse	.03	.09	.13	.05	.14
No response	.30	.27	.23	.21	.32
Fail*	.36	.38	.40	.28	.49
	(n = 5)	(n = 13)	(n = 9)	(n = 3)	(n = 10)
Rigid	.13	.10	.11	.14	.09
Modify	.37	.41	.33	.33	.44
Clarify	.00	.00	.03	.00	.00
Self solve	.01	.02	.03	.03	.02
No further response	.35	.31	.33	.41	.28
Change subject	.10	.05	.05	.03	.06
Follow partner	.00	.10	.11	.06	.11
	(n = 8)	(n = 20)	(n = 9)	(n = 10)	(n = 10)
Elicit action	.24	.35	.33	.36	.34
Stop action	.16	.19	.19	.20	.17
Acquire object	.03	.08	.06	.10	.05
Gain attention	.51	.31	.32	.26	.36
Joint action	.06	.08	.10	.08	.08

*Fail = acknowledge + refuse + no response.

children did not differ from their neutral counterparts on the production of requests. Similarly, the sociable children did not differ from their neutral playmates concerning the frequency with which different request types were emitted. However, it was discovered that neutral children paired with sociable partners produced significantly more direct requests, $t(18) = 2.37$, $p < .03$, and more requests of both types combined, $t(18) = 2.17$, $p < .04$, than did neutral children paired with isolate partners.

Consequences of Requests

The isolate children did not differ significantly from their neutral playmates concerning the consequences of their direct and indirect requests. Thus, despite the finding that 65 percent of all the requests by neutral children paired with isolates were successful, as contrasted with 54 percent for the isolates' requests, these differences were nonsignificant. It is possible that the small n's contributed, in large part, to this nonsignificant finding.

Significant differences were found when request consequences were compared between the sociable children and their neutral playmates. The proportions of direct, indirect, and total requests (direct and indirect) that led to no response from the partner were greater for neutral than sociable children, $t(18) = 2.10$, $p < .05$; $t(18) = 2.18$, $p < .02$; and $t(18) = 2.64$, $p < .02$, respectively. Trends were found for the proportion of all requests that reached a successful conclusion, $t(18) = 1.83$, $p < .09$, and for the proportion of all requests that resulted in failure, $t(18) = 1.72$, $p < .10$. Sociable children experienced slightly more success and less failure than did their neutral counterparts.

Finally, neutral children paired with isolates versus sociable children experienced different degrees of success following the production of requests. Neutral children paired with isolates (NI children) were more successful at gaining compliance with their direct requests than were neutral children paired with sociables (NS children), $t(18) = 2.56$, $p < .02$. NS children were more likely to experience failure, $t(18) = 3.30$, $p < .004$, and to receive no response to their direct requests, $t(18) = 2.42$, $p < .03$, than were NI children. Furthermore, NS children were more likely to have their partners refuse to comply with all their requests, direct and indirect, $t(18) = 2.32$, $p < .04$. They were also more likely to gain no response to their requests, $t(18) = 2.19$, $p < .04$, and to have their requests fail, $t(18) = 3.39$, $p < .003$, than NI children. On the other hand, NI children's requests (direct and indirect) were more likely to result in success than those of NS children, $t(18) = 2.44$, $p < .03$.

Strategies Following Failure

Isolate children were significantly more likely than NI children to modify their direct requests, $t(8) = 2.54$, $p < .04$, and less likely to modify their indirect requests following failure, $t(13) = 3.28$, $p < .006$. Sociable children were less inclined than NS children to modify their indirect requests, $t(18) = 3.54$, $p < .002$, and to modify all requests, $t(17) = 2.17$, $p < .04$, following failure. No other rerequest strategies were found to differentiate significantly these groups. NS children were also more likely to modify their original direct requests following failure than were NI children, t(13) = 2.34, p < .04.

Types of Direct and Indirect Requests

Isolate children emitted proportionately more direct, indirect, and total requests to gain attention, t(16) = 2.36, p < .03; t(17) = 2.08, p < .05; and t(16) = 2.44, p < .03, respectively, than did their NI playmates. Isolate youngsters produced significantly fewer direct requests to elicit action, t(16) = 2.10, p < .05; fewer total requests to elicit action, t(16) = 2.07, p < .05; fewer indirect requests to acquire objects, t(17) = 2.03, p < .04; and fewer total requests to acquire objects, t(16) = 2.84, p < .01, than did their NI counterparts.

Sociable children and their NS play partners did not emit different types of requests. However, NI children directed significantly more direct and total requests to acquire objects to their play partners than did NS children, t(18) = 2.14, p < .05 and t(18) = 2.10, p < .05, respectively.

DISCUSSION

Do preschool and kindergarten children who have been identified as social isolates evidence communicative repertoires that differ from their more sociable age-mates? Can an examination of children's discourse aid in the discovery of potential sources of social withdrawal in early childhood? These two general questions guided the research described in this chapter.

At first glance, there appear to be few differences in the communicative repertoires of young children varying with regard to sociability. Isolate children produced as many direct, indirect, and total requests as their neutral play partners. However, when the request behaviors of isolates are compared with (a) neutral children who have been paired with both isolate and sociable partners, and (b) with sociable children, it is quite clear that they speak less than their counterparts. Analyses of variance comparing the frequency of direct, indirect, and all requests across the three groups were significant for the first and last dependent measures, Fs(2, 37) = 3.33, p < .05 and 3.64, p < .05. Isolate children produced fewer direct and total requests than did both groups of more sociable children.

Given these results and those concerning the nonsignificant differences in request production between isolate children and their neutral partners, it appears that socially withdrawn children elicit fewer communicative overtures from their playmates than do more sociable children. One may argue that the neutral children paired with isolate children were themselves quite different from their neutral counterparts paired with sociable children. However, as part of the larger study, of which this report is but a small piece, it was first ascertained that the two neutral groups of children did not differ significantly from each other concerning chronological or mental age and performance on a variety of social cognitive development tests (Rubin, 1982a). In short, it seems that isolate children did elicit rates and types of requests (see discussion below) different from their sociable age-mates.

These initial results inform us that classroom observations during which the children were targeted as isolate, neutral, and sociable do generalize to the dyadic setting. That is, socially withdrawn children are less sociable not only in the classroom, but also in the highly constrained playroom used in this study. The results, however, provide few clues concerning the sources of social withdrawal.

An examination of the consequences of children's requests begins to fill in the picture. The differences between isolate children and their neutral partners in the success and failure rates of requests were not statistically significant. These non-significant results may have emanated from the low power of our analyses given the rather small number of children studied. In point of fact, the isolate children's requests were successful 54 percent of the time, whereas those of their partners were successful 65 percent of the time. Moreover, the isolate children failed 36 percent of the time, whereas the neutral children failed only 28 percent of the time.

These findings, although not statistically significant, do appear psychologically significant given the differences in the types of requests emitted by these two groups of children. Isolate children's requests were more often made up of attention bids. These bids for attention required only that the listener glance momentarily at the speaker in order for "success" to be coded. On the other hand, their neutral partners more often produced elicit action and object acquisition requests, both "high cost" bids in that they require active compliance, often involving movement across the playroom. Moreover, such high cost bids often require the partner to cease his or her own current activities in favor of complying with the speaker's request. Given these differences in the types of requests produced by isolate and NI children, the nonsignificant differences in success and failure rates is nevertheless impressive. One might fully expect the emission of high cost requests to be accompanied by higher probabilities of failure for the NI group.

Interestingly, when neutral children were paired with sociable play partners, a number of differences were found concerning request consequences. NS children's requests were more often met with no response than were those of their more sociable partners. Moreover, there was a trend indicating a greater failure rate for NS than for sociable children's requests. These results take on added meaning upon examination of the consequences of NI versus NS children's requests.

NI children's requests were generally more successful and less likely to fail than those of their NS counterparts. For the most part, the types of requests produced by these two groups did not differ. These findings suggest that children's self-confidence and perceptions of dominance are manifested during communicative interaction. At the bottom of the dominance hierarchy are children who have low self-concepts; generate low-cost, nonassertive attention-getting requests; and who give in to or comply with the requests of play partners significantly more often than do sociable children. These children are, of course, the isolates. At the top of the dominance hierarchy are the sociable, self-assured children who generate high cost requests and who do not feel compelled to comply with the requests of their lower-status partners.

Admittedly, the conclusions offered above are highly inferential. Data concerning the dominance status and the self-concepts of the three target groups are obviously required in further studies. However, the conclusions do make implicit sense.

The strategies used by the three groups following request failure are mixed at best and do not appear to be helpful in identifying correlates of sociability status. The basic problem is simply that the small sample of children we were able to examine statistically at this level of analysis was too small for meaningful results to emerge. For example, the numbers of NI and isolate children who experienced failure for both direct and indirect requests were only three and five, respectively.

In summary, our data suggest that an examination of discourse is helpful in identifying correlates with and potential causes of children's sociability status. Earlier findings have demonstrated that isolate children are less cognitively and socially mature (Rubin, 1982a), are perceived by their teachers to be more anxious in peer group settings (Rubin & Clark, in press) and are more deficient in leadership skills (Kohn & Rosman, 1972) than their more sociable age-mates. Given these facts, it is conceivable that this group is also less self-assured and lower in the dominance hierarchy than their more sociable age-mates. The communication data corpus examined here reinforces this conclusion. With increasing sociability, the amount of socially directed speech increased, the success rates of requests increased, and the "costliness" of the directives produced increased. In short, with increased sociability, communicative overtures that reflected assertiveness and self-assurance likewise increased.

Further data concerning dominance and self-concept status most definitely are required in order to fill in the picture we have begun to paint. These data will be gathered as part of a longitudinal study presently being conducted on the children who served as subjects in this study. Nevertheless, the data presented herein do suggest a possible intervention strategy for young, socially withdrawn children. If, indeed, isolate children are less mature cognitively and socially than their more sociable age-mates, if they are more anxious in peer group settings, and if they are deficient in leadership skills, it might be appropriate to provide them with confidence-building peer group experiences. Such experiences might emanate from interactions with play partners who are on a social and cognitive plane similar to their own—that is, younger children. Furman, Rahe, and Hartup (1979) have, in fact, paired isolate children with younger play partners in an intervention study. The rationale underlying their peer-pairing strategy was to provide isolate children with the opportunity to play a dominant role in a dyadic partnership, thereby increasing self-confidence, assertiveness, and ultimately sociability. Furman et al. found that this peer-pairing procedure was more effective in increasing social interaction than was the pairing of isolate children with age-mates.

Although the product of this peer-pairing procedure is quite clear (and optimistic), the process through which it succeeded remains somewhat of a mystery. Are the communicative overtures of isolates more assertive when these children are in the company of younger playmates? Do isolate children become less compliant (and more self-assured) when paired with younger partners? These and other questions directed toward discovering the causes of social withdrawal, and the ways through which peer pairing intervention procedures work, may best be answered in future studies of communicative competence.

REFERENCES

Asher, S. R., & Hymel, S. (1982). Children's social competence in peer relations: Sociometric and behavioral assessment. In J. D. Wine & M. D. Smye (Eds.), *Social competence*. New York: Guilford Press.

Botvin, G., & Murray, F. B. (1975). The efficacy of peer modeling and social conflict in the acquisition of conservation. *Child Development, 46*, 796-799.

Burns, S. M., & Brainerd, C. J. (1979). Effects of constructive and dramatic play on perspective-taking in very young children. *Developmental Psychology, 15*, 512-521.

Campbell, S. B., & Cluss, P. (1982). Peer relationships of young children with behavior problems. In K. H. Rubin & H. S. Ross (Eds.), *Peer relationships and social skills in childhood*. New York: Springer-Verlag.

Charlesworth, W., & Hartup, W. W. (1967). Positive social reinforcement in the nursery school group. *Child Development, 38*, 993-1003.

Cowen, E. L., Pederson, A., Babigian, H., Izzo, L. D., & Trost, M. A. (1973). Long-term follow-up of early detected vulnerable children. *Journal of Consulting and Clinical Psychology, 41*, 438-446.

Delia, J. G., Kline, S. L., & Burleson, B. (1979). The development of persuasive communication strategies in kindergarteners through twelfth-graders. *Communication Monographs, 46*, 241-256.

Furman, W., Rahe, D. F., & Hartup, W. W. (1979). Rehabilitation of socially withdrawn preschool children through mixed-age and same-age socialization. *Child Development, 50*, 915-922.

Garvey, C. (1975). Requests and responses in children's speech. *Journal of Child Language, 2*, 41-63.

Hartup, W. W. (1983). Peers as socialization agents. In E. M. Hetherington (Ed.), *Handbook of child psychology: Socialization, personality and social development*. New York: John Wiley.

Iannotti, R. (1978). Effects of role-taking experiences on role-taking, empathy, altruism, and aggression. *Developmental Psychology, 14*, 119-124.

Keller, M. F., & Carlson, P. M. (1974). The use of symbolic modeling to promote social skills in children with low levels of social responsiveness. *Child Development, 45*, 912-919.

Kohn, M., & Rosman, B. L. (1972). A social competence scale and symptom checklist for the preschool child: Factor dimensions, their cross-instrument generality, and longitudinal persistence. *Developmental Psychology, 6*, 445-452.

Krasnor, L. R., & Rubin, K. H. (1981). Social problem-solving skills in young children. In T. Merluzzi, C. Glass, & M. Genest (Eds.), *Cognitive assessment*. New York: Guilford Press.

Lamb, M. E., & Roopnarine, J. (1979). Peer influences on sex-role development in preschoolers. *Child Development, 50*, 1219-1222.

Levin, E., & Rubin, K. H. (1983). Getting others to do what you want them to do: The development of children's requestive strategies. In K. Nelson (Ed.), *Children's language: Vol. 4.* New York: Lawrence Erlbaum.

Miller, S. A., & Brownell, C. A. (1975). Peers, persuasion and Piaget: Dyadic interaction between conservers and non-conservers. *Child Development, 46,* 992-997.

Parten, M. (1932). Social participation among preschool children. *Journal of Abnormal and Social Psychology, 27,* 243-269.

Patterson, G. R., Littman, R. A., & Bricker, M. (1967). Assertive behavior in children: A step toward a theory of aggression. *Monographs of the Society for Research in Child Development, 32,* (5, Serial No. 113).

Piaget, J. (1926). *The language and thought of the child.* London: Routledge & Kegan Paul.

Piaget, J. (1932). *The moral judgment of the child.* New York: Free Press.

Roff, M., Sells, S. B., & Golden, M. M. (1972). *Social adjustment and personality development in children.* Minneapolis: University of Minnesota Press.

Rubin, K. H. (1979). The impact of the natural setting on private speech. In G. Zivin (Ed.), *The development of self-regulation through private speech.* New York: John Wiley.

Rubin, K. H. (1982). Social and social-cognitive developmental characteristics of young isolate, normal, and sociable children. In K. H. Rubin & H. S. Ross (Eds.), *Peer relationships and social skills in childhood.* New York: Springer-Verlag.

Rubin, K. H., & Clark, M. L. (in press). Preschool teachers' ratings of behavioural problems: Behavioural, sociometric and social-cognitive correlates. *Journal of Abnormal Child Psychology.*

Rubin, K. H., Fein, G. G., & Vandenberg, B. (1983). Play. In E. M. Hetherington (Ed.), *Handbook of child psychology: Socialization, personality and social development.* New York: John Wiley.

Rubin, K. H., & Pepler, D. J. (1980). The relationship of child's play to social-cognitive development. In H. Foot, T. Chapman, & J. Smith (Eds.), *Friendship and childhood relationships.* London: John Wiley.

Shaffer, D. (1979). *Social and personality development.* Monterey, CA: Wadsworth.

Spivack, G., & Shure, M. B. (1974). *Social adjustment in young children.* San Francisco: Jossey-Bass.

Suomi, S., & Harlow, H. (1972). Social rehabilitation of isolate-reared monkeys. *Developmental Psychology, 6,* 487-496.

PART III

SOCIAL COGNITIVE AND STRATEGIC PROCESSES IN ADULTS: THE ORGANIZATION OF COMMUNICATION

7

Individual Differences in Communication:
Social Cognitive Determinants and Consequences

C. Douglas McCann and E. Tory Higgins

Because verbal communication is distinctive to human social interaction, one would expect its determinants and consequences to be a central issue in personality and social psychology. Suprisingly, this has not been the case. One possible reason for this, perhaps, has been the absence of a framework from which to explore both the interpersonal aspect of communication and the role of individual differences in language use. Individual differences have received considerable attention in sociolinguistics, but sociolinguistic analyses are insufficient for our purposes. Thakerar, Giles, and Cheshire (in press) have voiced a number of reservations in this regard. First, sociolinguistic analyses often lack predictive power due to their tendency to be descriptive rather than explanatory. Second, sociolinguistic analyses tend to emphasize correlations between speech and "large-scale, objectively defined" variables (such as social class) and have largely ignored the role, for example, of participants' goals as determinants of communicative behavior. Third, such analyses have ignored communication as an independent variable. Finally, sociolinguistics has not adequately considered the relation between linguistic and nonlinguistic factors in terms of their relative importance in interpersonal interaction.

By taking a *social cognitive* perspective, we hope to eliminate some of these shortcomings. The social cognitive perspective involves a consideration of both the social or interpersonal determinants of information processing (the social psychology of cognition), and the cognitive mediators of social judgment and behavior (the cognition of social psychology). Previous research in communication and language use has reflected both of these aspects of social cognition. Reflecting the social psychology of cognition, one line of research (Higgins, 1981; Higgins, Fondacaro, & McCann, 1981) has demonstrated that momentary social factors can affect how people talk about their social world, and that this, in turn, can bias how social events are remembered and evaluated ("Saying is believing").

Reflecting the cognition of social psychology, another line of research (Higgins & King, 1981) has demonstrated that momentary and chronic individual differences in the accessiblity of different constructs can cause individual differences in the encoding and categorization of social stimuli. Our major purpose in this chapter is to expand on our earlier work by exploring the nature of individual difference in language use, and the ways in which such differences contribute to the subjective construction of social reality. We will first consider the game-like features of communicative interaction, and their implications for individual differences in communication goals and the selection of strategies for fulfilling these goals. We will then consider more unconscious, automatic determinants of language use, in particular social construct accessibility (Higgins & King, 1981). Finally, the implications of such social cognitive mediators for the interrelationship among language, thought, and society will be considered.

THE "COMMUNICATION GAME" AND INDIVIDUAL DIFFERENCES IN LANGUAGE USE

Communication is a shared interpersonal enterprise. It involves shared patterns of expectations, rules, or conventions concerning participants' social roles and appropriate language use (Austin, 1962; Cushman & Whiting, 1972; Gumperz & Hymes, 1972; Morris, 1964; Peirce, 1940; Rommetveit, 1974; Reusch & Bateson,

1968; Searle, 1976; Watzlawick, Beavin, & Jackson, 1967). Effective and efficient communication requires co-orientation and monitoring between participants, with social role-taking being a prerequisite skill (Cushman & Whiting, 1972; Delia, 1976; Grice, 1971; Mead, 1934; Piaget, 1926; Rommetveit, 1974; Searle, 1976). Communication serves not only to transmit information, but also to establish and maintain relationships between participants (Blumer, 1962; Garfinkel, 1967; Gumperz & Hymes, 1972; Hawes, 1973; Watzlawick et al., 1967). As a social phenomenon, it also establishes a context through which participants define the social meaning of the interchange (Blumer, 1962; Garfinkel, 1967; Goffman, 1959; Hawes, 1973; Rommetveit, 1974; Watzlawick et al., 1967).

Such prototypic features of interpersonal communication raise issues of obvious importance to social psychology. In this chapter we hope to demonstrate the heuristic value of establishing a social psychology of communication that integrates social psychology and linguistics. Previous social psychological treatments of communication were restricted largely to an examination of communication as "information transmission" (see Higgins, 1981; Higgins et al., 1981, for a more complete discussion of this point). Their focus was primarily on referential communication (for example, Glucksberg, Krauss, & Higgins, 1975; Mehrabian & Reed, 1968) and persuasive communication (for example, Eagly & Himmelfarb, 1978; McGuire, 1969, 1972). Although this approach has contributed greatly to our knowledge of certain aspects of communication, it has failed to highlight the truly social nature of interpersonal communication, in particular the "social relation" features of communication. For example, this approach has paid little attention to the cultural conventions for language use across different topics and social situations, to interpersonal or social goals other than information transmission (for example, "face" goals, social reality goals, and social relationship goals), or to distinctions between speaker and listener roles.

The present section will focus on a recent conceptualization of social communication that emphasizes its game-like features (for example, Clark & Delia, 1979; Garfinkel, 1967; Goffman, 1959; Lyman & Scott, 1970; Wittgenstein, 1953). The "communication game" approach (Higgins, 1981; Higgins et al., 1981) emphasizes normative prescriptions regarding appropriate communicative behavior and the goal-oriented nature of communicative interaction (McCann & Hancock, in press). Although these two aspects of interpersonal communication are interdependent, we will examine them separately.

Rules of Interpersonal Communication

The communication game approach focuses on social relations in interpersonal communication and reflects an elaboration and integration of approaches derived from the fields of speech communication, philosophy, linguistics, anthropology, and sociology. In this framework, communication its conceptualized as a game in the sense that it involves "purposeful social interaction occurring within a socially defined context, involving interdependent social roles and conventional rules, strategems, and tactics for making decisions and obtaining various goals" (Higgins, 1981, p. 346). A review of the literature suggests the following general rules of the communication game (Austin, 1962; Cushman & Whiting, 1972; Grice, 1971; Gumperz & Hymes, 1972; Rommetveit, 1974; van Dijk, 1977):
Communicators should

(1) take the audience's or recipient's characteristics into account;
(2) convey the truth as they see it;
(3) try to be understood (that is, be coherent and comprehensible);
(4) give neither too much nor too little information;
(5) be relevant (that is, stick to the point);
(6) produce a message that is appropriate to the context and circumstances;
(7) produce a message that is appropriate to their communicative intent or purpose; and
(8) assume that the message recipient is trying as much as possible to follow the general rules of communication.

Message recipients should

(1) take the communicator's or source's characteristics into account;
(2) determine the communicator's intent or purpose;
(3) take the context and circumstances into account;
(4) pay attention to the message and be prepared for receiving it;
(5) try to understand the message;
(6) provide feedback to the communicator concerning the recipient's interpretation or understanding of the message; and
(7) assume that the communicator is trying as much as possible to follow the general rules of communication.

This list is not meant to imply either that these are the only rules of verbal communication, or that people are always aware of these rules. The list is also not meant to suggest that all of these rules must be followed for communication to be successful or acceptable. In fact, in many cases it is not possible to follow all of the rules, and some must and will take precedence over others depending on the circumstances and purposes of the communicative interaction. Thus, unlike formal games, there is no explicit set of rules, all of which must be followed to play the game properly. As we will argue below, individual differences in goal emphasis play a large role in determining which rule will be followed in any particular communicative interaction.

Much of our work to date has centered on the implications of rule 1. A considerable amount of research and theory suggests that effective and efficient communication requires that the communicator modify his or her message by taking the audience's characteristics into account (Flavell, Botkin, Fry, Wright, & Jarvis, 1968; Glucksberg et al., 1975; Higgins, 1977; Higgins & Rholes, 1978; Manis, Cornell, & Moore, 1974; Mead, 1934; Newtson & Czerlinsky, 1974). Although past research has demonstrated such message modification, the communication game perspective also considers the unanticipated consequences of such message modification for the communicator.

Higgins and Rholes (1978) examined the implications of such rule-prescribed message modification by having subjects expect to deliver a referential message concerning a target person to a listener who supposedly either liked (positive audience condition) or disliked (negative audience condition) the target person. It was expected that subjects would modify their messages to suit their audience's attitude toward the target person. In addition, because of the effects of labeling and categorization on memory and judgment, it was expected that such message modification would cause subjects to remember and evaluate the target person later in a manner that was evaluatively consistent with their messages. The stimulus information concerning the target person contained evaluatively ambiguous items, unambiguously negative items, and unambiguously positive items.

Higgins and Rholes found support for the predicted rule-following behavior (that is, message modification) and for the subsequent effects of this behavior on subjects' reproductions of the original stimulus information and attitudes toward the target person. Subjects distorted their messages to be consistent with the attitudes of their listener, and subsequently recalled stimulus information in a manner that was evaluatively consistent with these distorted messages. In addition, subjects' own attitudes toward the target person reflected the evaluative tone of their messages; that is, they became more positive over time in the positive audience condition and more negative over time in the negative audience condition. Finally, the effects of message encoding on recall and attitudes increased over a two-week period. These results were interpreted in terms of subjects relying more and more on their verbal encoding of the initial stimulus for their subsequent recall and evaluative judgments of the stimulus as the details of the stimulus information were gradually forgotten (see Bartlett, 1932; Higgins, Rholes, & Jones, 1977).

Higgins and Rholes (1978) focused on message modification to suit the attitudinal characteristics of one's audience. A subsequent investigation (Higgins, McCann, & Fondacaro, 1982) examined message modification to suit the information needs of one's audience. Subjects in this study were assigned to the role of speaker or listener. Subjects in both roles were told that they would receive a stimulus essay describing another student at their university, but speakers expected to communicate their own impressions of the target person, while listeners expected to receive a speaker's impressions of the target. Each speaker and listener was told that his or her communicative partner had either the same information or different information. (In reality, all subjects received the same information.) After reading the essay, subjects were asked to reproduce the essay word for word.

It was expected that when their recipient had *different* information about the target person, speakers would emphasize description over interpretation in order to provide the recipient with a complete and accurate account of the target person. When the recipient had basically the *same* information about the target person, however, communicators were expected to emphasize interpretation over description so that their message would not be redundant from the recipient's perspective. This is what was predicted if communicators followed rule 4 and gave neither too much nor too little information. As anticipated, speakers varied their messages according-to the information needs of their listeners. Speakers were less likely to change stimulus information if they believed that their communicative partner had information different from their own. In addition, message modification had considerable impact on the speaker's own memory of the target person. Recall accuracy over time was dependent on the extent to which the speakers were careful to "stick to the facts" in their messages.

These studies examined the effects of rule-following on the speaker. The communication game, however, also has implications for the listener. Recipient rules 1 and 2 state that recipients must take the speaker's characteristics, motives, and intentions into account. A recipient's judgment of communicator characteristics will be based on cues available in the communicative interaction. If recipients know that the communicator is aware of their own attitude toward a particular message topic, for example, they may take this into account when judging the speaker's message. Research has shown that people adjust their view of a speaker's true opinion on the basis of the opinion of the speaker. That is, subjects will "correct" their judgment of the speaker's opinion away from that of the audience (Newtson & Czerlinsky, 1974). In addition, recipients are more likely to perceive that the attitude conveyed in a message accurately reflects the speaker's true position if either the expressed attitude is contrary to the recipient's attitude, is against the speaker's best interest, or if the recipient has "overheard" the message (Eagly, Wood, & Chaiken, 1978; Mills & Jellison, 1967; Walster & Festinger, 1962). In sum, recipients, like communicators, must and do take their partner's characteristics into account.

Although the communication game perspective emphasizes the similarities inherent in speaker and listener roles, it also focuses on the differences. Communicators, for example, should try to be understood and recipients should be prepared for receiving a message. Although these differences are necessary and complementary, they are likely to have differential consequences for individuals occupying these roles. Because communicators are expected to produce clear, concise messages, they should tend to polarize and distort stimulus person information more than recipients, who are supposed to be open to change and prepared for a wide range of possible information in the message (that is, to suspend judgment). This prediction has received strong support in the "cognitive tuning" literature (Cohen, 1961; Harkins, Harvey, Keithly, & Rich, 1977; Leventhal, 1962; Zajonc, 1960).

In addition to these role-produced differences, speakers and listeners in these studies have differed in terms of whether they anticipated additional information regarding the message topic. Listeners, but not speakers, expected additional information from their communicative partners regarding the target person or issue. This may have led to the polarization and distortion differences described above, since subjects as speakers would be more likely to integrate information into a unified impression if they believed that they had all the relevant information. Listeners, on the other hand, expected a message from their speakers and might therefore be predicted to suspend judgment.

Differences in expectations about receiving additional information, as well as the role-prescribed goals of communicators and recipients, were independently manipulated in the Higgins, McCann, et al. (1982) study. After subjects had been assigned their communicative role (speaker or listener), half were told that they would later receive additional information about the target person and half were not. No subject in either communicative role actually received additional information. Prior to any communicative interaction, all subjects were asked to write down their impressions of the target person. The impressions were scored by counting the number of positive and negative trait labels used by the subjects to refer to behavior contained in the stimulus essay. The result of this analysis indicated, in general, a strong positivity bias across all conditions.

Consistent with previous cognitive tuning studies, communicators who expected no additional information. Prior to any communicative interaction, all subjects were asked to write down their impressions of the target person. The impressions were scored by counting the number of positive and negative trait labels used by the subjects to refer to behavior contained in the stimulus essay. The result of this analysis indicated, in general, a strong positivity bias across all conditions.

Consistent with previous cognitive tuning studies, communicators who expected no additional information showed a positivity bias. The information reproduced by these subjects in their impressions tended to be more positive in implication than the original stimulus information they had read. Again consistent with the previous literature, recipients who expected additional information showed no evidence of such polarization in their impressions. Our primary interest, however, centered on two other conditions; that is, the communicators in the "expectation" condition and the recipients in the "no expectation" condition. Analysis showed that both of these groups evidenced a positivity bias, indicating that while communicators' positivity bias (polarization) is largely a function of their following a communicative rule (be concise and coherent), recipients' positivity bias can be determined by whether or not they expect to receive additional information.

As is clear from the above discussion, the communication game approach focuses on the effects of rule-following behavior in communicative interaction. However, this approach also emphasizes the role played by interaction goals in the communication process. Indeed, communication goals are assumed to be largely responsible for inducing individuals to emphasize or follow some communication rules rather than others. We turn now to a discussion of this component of the communication game.

Goals of Interpersonal Communication

In contemporary social cognition research, there has been an increasing emphasis on the role played by personal goals in processing and encoding social information (Cohen, 1981; Cohen & Ebbesen, 1979; Hamilton, Katz, & Leirer, 1980; Higgins, McCann, et al., 1982; Jeffery & Mischel, 1979; McCann, 1982a,; McCann & Hancock, in press). A key feature of the communication game is its emphasis on the goal-oriented nature of communicative interactions.

Individuals enter into communicative interactions in order to achieve a variety of personal goals. These goals have been classified by Higgins, Fondacaro, et al. (1981) as providing either intrinsic or extrinsic reasons for engaging in a communicative act. A variety of extrinsic goals have been identified. Beginning with Festinger's (1950, 1954) classic work on social comparison, social psychologists have been interested in how group interaction helps people to arrive at a consensually validated definition of "social reality." If this goal is blocked, communicative activity should increase in an effort to accomplish this end. In addition to social reality goals, interpersonal communication also serves to initiate, define, and maintain social relationships (Rusch & Bateson, 1968; Watzlawick et al., 1967). Social communication can also be used in the service of "identity" goals (Clark & Delia, 1979). Recent work on speech accommodation (Giles & Smith, 1979; Thakerar et al., in press) suggests that language is used as a tool for establishing and maintaining group-membership identity. When people's language group identity is threatened by a member of an outgroup, for example, they often diverge in language use in order to assert their ethnic identity (Ryan, 1979).

Communication often serves "face" (Goffman, 1967) or ingratiation goals (Jones, 1964). It is also initiated in pursuit of task or "group locomotion" goals (Festinger, 1950) in which communicative interaction is a necessary prerequisite to successful group problem solving. Social psychologists have been intrigued by the natural antagonism between task- and relationship-oriented social motives (Bales, 1955; Fiedler, 1971), a dichotomy that anticipates our later consideration of goal conflict in communicative interactions.

Everyday observation also indicates that people communicate to pursue intrinsic goals (Higgins, Fondacaro, et al., 1981), the most important of which are entertainment goals (Tubbs & Moss, 1977). It often seems that we approach particular others because they are "fun to talk to." Although entertainment is an important function of communication, this type of interaction goal has received little attention. This is particularly surprising given the importance attributed to affect in other areas (for example, Zajonc, 1980).

Communicative interactions, therefore, are initiated in pursuit of a variety of personal and interpersonal goals (see also Clark and Delia's, 1979, discussion of communicative objectives). These goals are important for two reasons. First, they systematically affect how people process and utilize the social information available in an interaction (Cohen, 1981; Jones & Thibaut, 1958). Second, they allow for an examination of individual differences in the emphasis given to different goals (McCann & Hancock, in press), including developmental differences (Garvey & Hogan, 1973; Higgins, Fondacaro, & McCann, 1981). If reliable individual differences can be related to these goals, our understanding of the antecedents and consequences of communicative behavior will be increased.

An emphasis on communication goals also highlights the goal conflicts and decision making involved in communication (Higgins, Fondacaro, & McCann, 1981; O'Keefe & Delia, 1982). Since people communicate to achieve a wide variety of goals, and since all goals cannot be achieved in any particular interaction, several types of goal conflicts can occur. Participants can differ in the means they choose in the pursuit of a common goal. Thus, even when participants are "promotively interdependent" (Deutsch, 1962), a form of social conflict may result. Different participants may be trying to achieve different goals in an interaction, which may result in conflict (for example, task versus social relationship goals). Finally, each individual typically approaches a social interaction with several goals in mind. In most cases, it will not be possible to satisfy all goals; thus, he or she must decide how much emphasis to give each goal.

The forces determining individual communicative behavior derive from both culture and personality. Communicative behavior is part of one's cultural system (Bernstein, 1970; Cazden, 1970; Higgins, 1976; Scribner & Cole, in press). Hence, the relative emphasis given different communication goals is partially determined by one's language community. Much of the sociolinguistic literature describes and analyzes the influence of such social variables on language and communication (for example, Trudgill, 1975). Role obligations can also influence communication. Cazden, Cox, Dickinson, Steinberg, and Stone (1979), for example, suggest that teaching emphasizes both interpersonal and task goals. While it is clear that leadership requires a shifting emphasis on social-emotional and task-related goals (Bales, 1955), individual factors are also important. Indeed, chronic individual differences in the emphasis placed on these goals are central to some theories of leadership (for example, Fiedler, 1971).

In summary, we are suggesting that communicative interactions are goal-oriented in nature and that these goals may affect communicative activity through their impact on which communication rules are followed. Which goals are given priority is determined by a variety of forces, including cultural conventions and personality-related needs. In addition to providing a useful framework for future research, this orientation suggests some new ways of interpreting previous research on individual differences in language use. We turn now to a consideration of both of these potential applications of the communication game perspective.

INDIVIDUAL DIFFERENCES IN GOAL ORIENTATION

It is our position that the needs or goals of participants determine the emphasis they put on different communication rules. These goals may derive from personality attributes or from larger, sociocultural variables such as social class, minority group membership, and sex. Although other fields have long been concerned with such variables, the communication game approach provides a new perspective on them. The value of the present framework derives in part from its ability to synthesize this diverse work. We will not review these literatures, but rather discuss issues that we feel are amenable to a communication game analysis.

Social Class Differences in Goal Orientation

Psychologists, anthropologists, and particularly sociologists have examined social class differences in language use and communication. This research has taken many forms. Strauss and Schatzman (1955), for example, found that lower-class respondents exhibited a pattern of straying from interview topics more often than respondents of other social classes. Johnson (1977) found that working-class children, when asked to encode abstract stimulus items in a referential communication task, tended to make more use of a "part-description" coding style than middle-class children, who in turn made more use of a "whole-inferential" style (see also Heider, 1971). Although most of this research has been conducted in Western cultures, some research shows that many social differences are constant across cultures (Straus, 1968).

Perhaps the most provocative social class studies have been those testing the ideas proposed by Bernstein (1960, 1970) on social class differences in the use of "restricted" and "elaborated" codes (Cazden, 1968; Hawkins, 1969; Higgins, 1976; Labov, 1970; Trudgill, 1975; Turner & Pickvance, 1971; Warren, 1966). From his analysis of narratives, Bernstein (1960) described the speech of middle- and working-class children as conveying relatively "universalistic" and "particularistic" meanings, respectively. According to Bernstein, the meanings conveyed by the universalistic speech forms of the middle-class children were context-independent; that is, their meaning depended on few features particular to the communication context. The particularistic speech forms of working-class children, on the other hand, were closely tied to communication contexts, in that their meaning required knowledge of the context. Bernstein (1970) asserted that individuals have available to them two potential speech forms, the elaborated and restricted codes, and that these forms are generally associated with the middle and working class, respectively. According to Bernstein, the association of these forms with social classes occurs through socialization processes. Thus, middle-class mothers convey to their children context-independent or universalistic meanings, while working-class mothers convey meanings more relevant to particular acts than normative considerations. Working-class speech emphasizes group membership and reference, with its concomitant linguistic dependence on shared contexts and linguistic cues.

Bernstein's interpretations of social class differences in speech code are problematic given the assessment procedures typical of some of this research (for example, Higgins, 1976; Labov, 1970). In a typical assessment situation, a child may be interviewed by a middle-class adult who attempts to determine the child's linguistic ability. This situation would create anxiety even in middle-class children who shared the interviewer's social class. For working-class children, the situation was even more stressful because of social class differences between the children and the interviewers.

It is likely that for working-class children the assessment situation was more than an achievement context, because they were faced with an outgroup member in a foreign context. As a result, the children might have adopted goals inappropriate from the interviewer's perspective. Finding themselves in a situation for which their language community had provided no normative prescriptions, the children might have (a) guessed (often incorrectly) which goals and associated rules were appropriate; or (b) applied inappropriate rules in an effort to achieve inappropriate goals (in terms of the performance objectives of the interviewer). If the performance context were changed, so would the result of the assessment. Indeed, Labov (1970) reports improvement in the linguistic performance of "disadvantaged" children when the assessment procedure was changed appropriately.

Individuals from different social classes differ in their familiarity with particular rules of communication and the situations to which they apply. They may differ as well in how they define the same social context. Working-class children may communicate appropriately according to their own, culturally based definition of the situation, but inappropriately according to the interviewer's definition of the situation. If a single definition were ensured, social class differences in performance might be attenuated. It is possible, however, that faced with a strange outgroup member and/or a situation to which they were not accustomed, working-class children would persist in their own goal orientation. In pursuit of identity goals, for example, individuals may attempt speech "divergence" (Thakerar et al., in press) in order to protect themselves from a threat to their group identity.

This goal-oriented interpretation of social class differences has interesting implications. It suggests the need for systematic research on social class differences in rule knowledge and goal orientation. Although little research has addressed this issue, some evidence suggests that different cultural groups do emphasize different goals (for example, Cazden, John, & Hymes, 1972). The present orientation also suggests that social class differences reflect performance rather than ability differences. Such performance differences, however, may have serious consequences. To the extent that different speech forms interfere with the co-orienting and communication monitoring necessary for effective and efficient communication, social class differences may lead to communication breakdown between participants from different social classes. Others have demonstrated the potential negative consequences of style discrepancies (for example, Tannen, 1981a, 1981b). The negative responses to such interchanges could, in turn, affect participants' memory for the content of the exchange and would surely affect their willingness to approach similar interactions in the future.

Minority Group Goal Orientation

Social psychologists have long been interested in studying intergroup relations (Amier, 1969; Sherif, Harvey, White, Hoo, & Sherif, 1961; Tajfel, 1974, 1979) and the implications of group membership for a wide variety of individual processes (Lewis, 1944; Lewis & Franklin, 1944). Group differences in language constitute an important aspect of intergroup relations and group membership (Taylor, Bassili, & Aboud, 1973; Taylor, Simard & Aboud, 1972). Lambert, Hodgson, Gardner, and Fillenbaum (1960), for example, have shown how easily language differences can elicit stereotypic perceptions of distinctive language groups by both outgroup and ingroup members. Given that different languages have different prestige, why do speakers of a lower-prestige language or dialect cling to that linguistic style (Ryan, 1979), despite the fact that it may hinder achievement, and may result in the adoption of negative stereotypes on the part of social perceivers?

Williams (1979) suggests that language group alignment is the result of rational decision making based on an individual's estimate of the utility of different courses of action. From this view, language alignment results from two processes, utility maximization and risk minimization. In order to better understand language group alignment, we must consider the conditions that lead to it.

Recent research suggests that distinctive minority group alignment often results from identity threats (Bourhis & Giles, 1977; Bourhis, Giles, Leyens, & Tajfel, 1979). Consider, for example, the research presented by Bourhis et al. (1979). The subjects were Flemish students who were to interact with a Francophone outgroup member in a language lab the students were attending to improve their English. In Belgium, there is considerable friction between the Flemish (members of the minority language group) and the French (members of the majority language group), and language is an important political issue. Students in the language lab were faced with a Francophone who was either sympathetic or unsympathetic toward the Flemish language. In addition, for half of the subjects interethnic group relations were made salient, and for half they were not. Subjects then responded to a series of tape-recorded questions from the outgroup member, who began to criticize the use of Flemish in Brussels and to argue for the use of French.

The results indicated increasing divergence (switching from the required English to Flemish) as Flemish was portrayed negatively, if the outgroup member was unsympathetic and intergroup relations had been made salient. It would appear, therefore, that this language criticism was perceived as threatening and aggressive, and that the Flemish speakers switched to their preferred language in order to strengthen their ethnic identity (that is, their identity goal).

Sex Differences in Goal Orientation

The literature on sex differences in language and communication is voluminous (for reviews, see Eakins & Eakins, 1978; Key, 1975; Philips, 1980; Thorne & Henley, 1975). We will not review this literature, but rather focus on an important and provocative approach to sex differences in language use and communication, specifically the work by Lakoff (1973, 1975, 1979). Lakoff proposes that women use some vocabulary men don't, employ more "empty adjectives" than men (for example, "sweet" and "divine"), and use questions where men use declarations. Especially notable is women's greater use of tag questions, which are considered to be midway between questions and declarations (for example, "We are going to the show, aren't we?"), hedges that are taken to convey uncertainty ("kinda" and "I guess"), intensifiers (so), and polite forms. Women also tend to err in the direction of overly correct speech forms. Although some have questioned the extent of these differences (for example, Eakins & Eakins, 1978; Philips, 1980), Lakoff's position seems very widely held.

Given that these sex differences are valid and reliable, what are we to make of them? How are we to understand and interpret these differences, especially given the orientation outlined in the present chapter? Perhaps the key lies in the realization that communicative behavior is a vehicle for the pursuit of various goals (for example, task, face, and social relationship goals), which may receive varying degrees of emphasis. Thus, these linguistic differences may actually reflect sex differences in goals. This interpretation is consistent with Philips's (1980, p. 54) suggestion that "the same features that sometimes differentiate males and females also serve other social functions as well."

What goals might these sex differences reflect? Given sexual stereotypes, the most obvious candidates are that male speech reflects task goals, while female speech reflects social relationship goals. These differing goals could be the result of several factors, such as educational and family socialization practices. For example, in a survey of children's textbooks used in schools, Wirtenberg and Nakamura (1976) found that men are frequently represented as the working (that is, "task-oriented") member of the couple. Even when women do work, their occupation is often portrayed as being other-oriented (teachers or nurses, for example). In addition, parents may teach traditional sex values in a variety of ways, often by generally encouraging more achievement orientation in males than females, at least on the part of fathers (Block, 1975).

Conklin (1974) argues that much of the work on sex differences indicates that women have a more acute sociolinguistic sensitivity; that is, more sensitivity to the states of others. Similarly, Barron (1971) suggests that women's speech indicates a "concern with the internal states of others." Gleser, Gottschalk, and Watkins (1959) found that females used more words that implied emotion and feeling. Wood (1966) found that males and females had distinct styles, with males being more task-oriented and females more socioemotionally oriented. Eakins and Eakins (1978) suggest that male-female differences reflect two distinct value systems. Masculine, "power-oriented" speech promotes individual supremacy and a competitive edge, while feminine speech reflects a concern with the welfare of the group and the importance of who one is rather than what one has accomplished. This research, although conducted in a variety of research traditions, converges in providing support for the sex differences in goal orientation suggested above. But perhaps the most comprehensive treatment of these differences is by Lakoff (1979).

Lakoff argues that, in a manner similar to the workings of generative grammar and transformational rules (Chomsky, 1965, 1968), specific rules govern and are predictive of linguistic styles (that is, there exists a "grammar of style"). In communicative interactions, speakers present themselves in ways prescribed by cultural and subcultural norms. Speakers' strategies fall on a continuum ranging from "clarity" (involving the least relationship between participants) through "distance" and "deference" to "camaraderie" (involving the most relationship between participants). "Clarity" refers to what we have called "information transmission" or "task" goals, while "camaraderie" is similar to what we have called "social relationship" goals. Within this system, idealized masculine and feminine roles can be represented by the clarity-deference and deference-camaraderie axes, respectively.

The similarity between our viewpoint and Lakoff's is striking. Both point to a sex difference in communication game goals in typical interactions. The primary goal of any interaction can both channel information search and determine interpretations of that information (Cohen, 1981; Jones & Thibaut, 1958). Goal differences characteristically exhibited by males and females may, therefore, have important implications for both the participants of any particular interaction and the course followed by the interaction itself. They may affect the perceptions of the participant by others and, in some cases, may strengthen existing sexual stereotypes. In addition, differing goals, through their effects on information search, attention, and interpretation, may affect participants' memory for the interaction. To the extent that only one set of objectives is reached, individuals with different objectives will be differentialy satisfied. This could lead to differences in the evaluation of an interaction.

Personal Goal Orientation

The communication game approach is concerned not only with the determinants and consequences of rule-following, but also with the role of participants' personal goals in the communication process. Given the importance of communication game rules for communicative behavior, a consideration of participant goals that could affect adherence to these rules should increase the predictability of communicative behavior.

An important type of goal that can be accomplished through communication is the face goal. In a recent study, one of us examined the implications of individual differences in the emphasis given to face goals (McCann & Hancock, in press). The social psychological construct of self-monitoring (Snyder, 1974, 1979) is relevant to our consideration of these goals. Out of a concern for the social appropriateness of their behavior, high self-monitors are especially attentive to cues from others in their environment as to how they should behave. They use these social cues to guide their behavior in particular social contexts. Low self-monitors, on the other hand, appear to act in accordance with their inner beliefs, attitudes, and feelings. One rule of interpersonal communication is that communicators should modify their message to suit the characteristics of their listener. Given the greater emphasis placed on face goals by high self-monitors, one would expect that message modification would be more characteristic of high than low self-monitors. High self-monitors are not only more sensitive to such audience information, but also tend to utilize it more in service of their strategic self-presentations.

It was suggested earlier that such message modification has unintended effects on communicators' memory and judgments of stimulus information. What would be the effects of such message modification for high self-monitors? Snyder (1979) suggests that high self-monitors realize that what they do and what they believe are not necessarily the same thing. Thus, their behavior may have little effect on their beliefs, and vice versa. On the other hand, the information processing mechanisms hypothesized to mediate the effects of the communication game would lead one to predict that high self-monitors would show effects similar to those described by Higgins and Rholes (1978). A recent study by McCann and Hancock (in press) was designed to address these issues.

In this study, each subject was introduced to an experimental confederate who was described as being a member of a group of students whose personalities and interpersonal relations had been previously assessed. The subjects were instructed that their task would be to describe another member of that group to the confederate so that he could try to identify him. All subjects were given an essay describing the target person and were told in an off-hand manner that previous observation had indicated that the confederate either liked (positive audience condition) or disliked (negative audience condition) the target person. Subjects were then instructed to write a message describing the target for the confederate. After a delay period, subjects reproduced the original descriptive essay and gave their impressions of the target person. Approximately ten days later, they came back to the laboratory and completed the dependent measures a second time.

The audience effects found by Higgins and Rholes (1978) were largely replicated. For example, an analysis of the evaluative distortions in subjects' messages revealed a significant interaction between audience condition and type of distortion, such that more positive distortion was found in the positive than the negative audience condition, and more negative distortion was found in the negative than the positive audience condition. Of particular interest in the present context, however, are the effects obtained for the self-monitoring comparisons. As expected, there was a significant interaction between audience type and the number of positive and negative labels for high self-monitors. High self-monitors used more positive labels in the positive than negative audience condition and more negative labels in the negative than positive audience condition. In contrast, the label analysis for low self-monitors did not reveal any comparable interaction.

It was also anticipated that such message modification would have consequences for subjects' subsequent impressions of the target person. As expected, there was an interaction between self-monitoring and audience condition, reflecting the fact that only the high self-monitors formed impressions that were consistent with their message labeling. A similar pattern of results was found for an analysis of the evaluative distortions contained in subjects' impressions. Only for high self-monitors was there an increase over time in the amount of impression distortions that were evaluatively consistent with message labeling. Finally, an analysis of the distortions contained in subjects' reproductions revealed a marginally significant interaction that was also consistent with the individual difference predictions. Once again, it was only the high self-monitors who distorted their reproductions of the original stimulus information in a manner that was evaluatively consistent with their message labeling.

Personal goals other than face goals can also influence communicative behavior. Participants in a communicative interaction, for example, establish a social relationship by engaging in communication (Rommetveit, 1974). One would therefore expect communicators to be influenced by their relationship to their partner when formulating and producing a message, but the impact of this factor should be dependent on the communication goals emphasized by the communicator.

One important relational dimension along which communicative participants vary is their social power, and "equal versus unequal" has long been recognized as a basic dimension of dyadic relationships (for example, French & Raven, 1959; Kelley, 1979). The importance of this power or status dimension varies for different people. In particular, there is a substantial literature indicating that the behavior of high authoritarians varies as a function of their partner's status to a greater extent than does the behavior of low authoritarians (Adorno, Frenkel-Brunswick, Levinson, & Sanford, 1950; Berg & Vidmar, 1975; Harvey & Beverly, 1961; Thibaut & Riecken, 1955). High authoritarians are more likely than low authoritarians to defer to and be influenced by those higher in status, especially for relationships involving authority. Thus, for dyadic communication involving a status or power differential, high authoritarians should emphasize social relationship goals more than low authoritarians when communicating to a high-status listener.

What would happen, for example, if communicators had to describe a target person for a listener who either liked or disliked the target person when the status of the listener was either higher than or equal to the communicator? We would expect that high authoritarians would modify their message toward their listener's attitude to a greater extent than would low authoritarians when the listener had higher status, but not when the listener had equal status.

In a recent test of this prediction (Higgins, McCann, & Fondacaro, 1982), we found this basic pattern of results on a measure of the relative amount of positive versus negative distortion of the original stimulus information contained in communicators' messages. Thus, high authoritarians do appear to emphasize social relationship goals over descriptive accuracy to a greater extent than do low authoritarians when their listener has high status. Consistent with the results of our previous studies (for example, Higgins & Rholes, 1978), we also found that communicators' messages influenced their own subsequent attitudes, impressions, and memory of the target person. In fact, communicators' impressions of the target person became increasingly consistent with their message over a two-week period. Interestingly, this impact of message modification on social cognition was more a function of the impact of the listener's attitude on the message than of the communicator's authoritarianism. Apparently, communicators were more likely to take into account (and adjust for) the impact of their personality on a message than the impact of the social circumstances (that is, their listener's attitude) on the message. Perhaps the social circumstances had a relatively unconscious impact on message formulation involving an automatic, rule-following process (following communicator rule 1), whereas personality had a relatively conscious impact on message output involving an active, goal-directed process (such as high authoritarians' desire to defer to high-status listeners). The results would then suggest that the communicators only took into account and adjusted for factors that they were aware had influenced their message.

Summary

In this section, we introduced an individual-difference perspective on interpersonal communication. This perspective was based primarily on the communication game model of social communication (Higgins, 1981; Higgins, Fondacaro, & McCann, 1981). According to this model, speakers and listeners follow general communication rules that govern their performance. Not all of these rules can be followed at the same time, however. An individual's choice of which rule to follow is dictated by sociocultural, contextual, and personal forces. Our analysis concentrated on the goal determinants of rule-following.

CONSTRUCT ACCESSIBILITY AND INDIVIDUAL DIFFERENCES IN LANGUAGE USE

In the previous section, we reviewed aspects of communication and language use that arise from active decisions and involve the fulfillment of goals and rules. But communication may also involve processes that are relatively unconscious and automatic. One such process concerns construct accessibility, which refers to the readiness with which a stored construct is utilized in information processing (Higgins & King, 1981). In this section, we consider the general hypothesis that individual differences in construct accessibility cause differences in the categorization and encoding of stimuli, which in turn have consequences for the process of interpersonal communication.

According to Bruner (1957), expectations and motivation are primary determinants of construct accessibility. The recent activation of a category has also been shown to influence its accessibility temporarily (Forbach, Stanners, & Hochhaus, 1974; Higgins, Rholes, & Jones, 1977; Huttenlocher & Higgins, 1971; Warren, 1972; Wyer & Srull, 1980). Especially pertinent to the present chapter are the effects of frequency of activation on a category's subsequent accessibility (Wyer & Srull, 1980). To the extent that frequent and repeated activation of a category is dispersed over a long time, the accessibility of the category will be increased for a prolonged period (Higgins & King, 1981). Thus, just as different socialization experiences influence the use of particular linguistic codes (Bernstein, 1970), and different linguistic systems affect which aspects of experience can be easily represented (Brown & Lenneberg, 1954), variation in socialization and linguistic experience might lead to relatively enduring individual differences in construct accessibility.

Much of the work on construct accessibility has considered how momentary, context-induced, individual differences in construct accessibility affect the verbal encoding and categorization of stimuli (Carmichael, Hagan, & Walter, 1932; Glucksberg & Weisberg, 1966; Higgins, Rholes, & Jones, 1977; Thomas & DeCapito, 1966; Wyer & Srull, 1980). This research clearly documents the importance of momentary differences in construct accessibility for social judgment and recall. It demonstrates that differences in the verbal categorization of stimulus input can result from momentary differences in construct accessibility. However, the focus here is on the consequences of chronic individual differences in construct accessibility, beginning with a consideration of the traditional language and thought literature from this perspective.

"Codability" and Its Cognitive Consequences

Whorf (Carroll, 1956) suggested that linguistic variables exert a powerful impact on cognition and thought. Fishman (1960) systematized the Whorfian hypothesis, distinguishing between the effects of lexical and grammatical influences on the world-view and behavior of linguistic groups. When testing Whorf's hypothesis, two research avenues are available. One could isolate cultural groups whose languages differ and then look for non-linguistic correlates ("between-language" approach). Alternatively, one could examine linguistic distinctions within a language and then examine nonlinguistic differences (a "within-language" approach).

Using a within-language approach, Brown and Lenneberg (1954) focused on the relation between the codability of different colors and subsequent memory for those colors. They examined several specific indices of codability and, after finding a high degree of relation among them, settled on the degree of intersubject agreement in color naming as their index of codability. Subjects in their experiment were first exposed to a series of color chips. After a brief delay, memory for the colors previously presented was tested using a recognition task that included a number of distractors. Their interest was in relating their previously obtained indices of codability to memory performance. Their results indicated a high level of association between codability and recognition, which increased as the encoding task became more difficult.

Subsequent work on codability and memory has produced mixed results (Heider, 1972; Lantz & Stefflre, 1964; Lucy & Schweder, 1981). However, it generally demonstrates a strong relation between linguistic and cognitive variables, suggesting that codability has important cognitive consequences. The Sapir-Whorf proposal can be reinterpreted within the context of our social cognitive framework. That is, different cultural goals and experiences can lead to chronic differences in construct accessibility that influence the encoding of stimuli. This in turn has important cognitive consequences for the encoder. It is possible that such codability and memory effects are the result of differential construct accessibility. In the traditional language and thought literature, researchers have found evidence for a relation between codability (categorization) and recognition memory, and as we have noted earlier, construct accessibility influences both categorization and memory.

This reasoning is consistent with traditional interpretations of this research. Brown (1965) and Lenneberg (1971), for example, proposed that the efficiency and effectiveness with which a linguistic system can represent a particular domain should influence the memorability of stimuli from that domain. It was thought that particular semantic categories werre more or less available to speakers of all languages. Most researchers seemed to suggest that it wasn't the absolute availability of those categories, but rather their relative availability (in other words, their accessibility). This framework suggests a more idiographic approach to the issue of the relation between language, thought, and society, an approach to which we will now turn.

Categorization Effects of Chronic Differences in Construct Accessibility

Many factors related to linguistic experience affect the accessibility of stored constructs. If individuals have different experiences, then they may also be expected to exhibit chronic individual differences in construct accessibility. Indeed, this appears to be part of the process tapped by early language and thought studies, especially when one considers the types of codability indices used in that research (for example, response latency and communication accuracy). The relevance of this to communication lies in the fact that: (a) category accessibility is a function of numerous factors that may have their roots in linguistic experiences (such as the frequency and recency of activation); (b) to the extent that such chronic differences affect various stages of the processing of incoming information, subsequent communicative activity will also be affected (O'Keefe & Delia, 1982); and (c) there is a close correspondence between category accessibility and labeling (Higgins & King, 1981).

Social stimuli are often ambiguous and can be represented in various ways. One factor that affects how a stimulus is encoded is the relative accessibility of the alternative categories that can be used to characterize it. Since conventional forms are dominant in interpersonal communication, it is natural to ask to what extent conventional forms are important in intrapersonal communication—that is, in memory (Lucy & Schweder, 1981, p. 147). The hypothesized importance of such linguistic variations is also evident in the verbal loop hypothesis of Glanzer and Clark (1962). They suggest that individuals encode experience into language and then store this linguistic abstraction for later use. If differences in construct accessibility affect this linguistic encoding, different representations of the experience may subsequently be recalled. Differential construct accessibility may even have implications for the many demonstrations of individual differences in language use (Dunn, Blis, & Siipilo, 1958; Foley & Macmillan, 1943; Jenkins, 1960; Nunally, 1965), although little research has been directed toward this issue.

Unfortunately, there has been little consideration to date of chronic differences in construct accessibility. Given differences in linguistic experiences, goals, expectations, and so on, we would expect similar differences in the accessibility of social constructs. There is some evidence for such differences, but few studies have addressed this issue directly (Beach & Wertheimer, 1961; Dornbusch, Hastorf, Richardson, Muzzy, & Vreeland, 1965; Kelly, 1955; Markus, 1977; O'Keefe, Delia, & O'Keefe, 1977; Yarrow & Campbell, 1963).

In two classic studies, Dornbusch et al. (1965) and Yarrow & Campbell (1963) had children describe other children with whom they were acquainted. The results showed that there was more construct overlap in descriptions involving a common perceiver and different targets than in descriptions involving two perceivers and a common target. However, this research provided only limited support for the hypothesized link between social judgment and differences in construct accessibility, because the measure of individual differences was based on the same data as the judgments (see Higgins, King, & Mavin, 1982, for a discussion of other problems). O'Keefe et al. (1977) did obtain independent measures of construct accessibility and target judgments, and their results suggest a relation between construct accessibility and social judgment. Unfortunately, that study did not control the stimulus information available to subjects and provides no direct evidence that construct accessibility affected the actual processing of information (see Higgins, King, et al., 1982).

In an earlier study, Thomas and De Capito (1966) examined the implications of individual differences in labeling. They identified subjects who spontaneously labeled an ambiguous color either green or blue. In a subsequent stimulus generalization test, they were told to respond whenever they recognized that color. The results indicated that subjects made more false recognition errors for colors resembling the category designated by their labeling of the ambiguous color than the other color, which suggests a relation between recognition memory and the chronic accessibility of different color terms. Glanzer and Clark (1962) examined the relation between the length of subjects' spontaneous verbal description of stimulus patterns and their subsequent recall of the arrays. They found not only a high correlation between the two measures, but consistent differences in the types of verbalization used by subjects.

Unfortunately, all of these studies suffer from one or both of two problems. First, several of the studies fail to utilize independent measures of construct accessiblity and memory, or to control the stimulus information available to subjects. Second, in some studies individual differences in perception could yield both verbal encoding and memory results. Thus, differences in the accessibility of categories may not have been independent of differences in the perception of stimuli.

Higgins, King, and Mavin (1982) attempted to overcome these limitations. In one study, undergraduates participated in two ostensibly unrelated experiments. In the first phase, subjects described two male friends, two female friends, and themselves. Accessible traits were those that appeared in the self-description of the subject and that of at least one friend, or those that appeared in the description of at least one friend, or those that appeared in the descriptions of at least three friends.

Two weeks later, all subjects were given the *same* paragraph to read, which contained behavioral descriptions of five positive and five negative traits that had been selected from those found to be accessible for some but not all subjects in the first phase of the study. By this procedure, some of the behavioral descriptions in the essay exemplified accessible traits, while the remainder exemplified inaccessible traits. After a brief delay, subjects reproduced the essay and gave their impression of the target person. Two weeks after this, subjects returned supposedly for another part of the study and were again given the reproduction and impression measures. Reproductions and impressions were scored for the number of accessible and inaccessible traits that were omitted. As expected, the results revealed that more inaccessible than accessible traits were omitted both immediately and after two weeks. These results suggest that variability in construct accessibility can affect memory for and impressions of stimulus persons.

In a second study, subjects again expected to participate in two unrelated studies. In the first phase of the study, accessible constructs were elicited by asking subjects to list traits associated with (a) a type of person that they liked, (b) a type of person that they disliked, (c) a type of person that they sought out, (d) a type of person that they avoided, and (e) a type of person that they frequently encountered. In the second session, subjects read individually tailored essays describing a person who possessed six accessible traits and six inaccessible traits. For each subject, accessible descriptions exemplified traits that were the first to come to mind (that is, listed first) in describing each of the persons mentioned above. Inaccessible traits were taken from the accessible

traits of other subjects so that a "quasi-yoking" procedure was used. In addition, inaccessible traits were selected to ensure that,
across all subjects, the accessible traits of one subject were used
as inaccessible traits for another, thereby controlling for essay content. After a delay, subjects reproduced what they had read and
gave their impressions of the target person. As predicted, analysis
of both reproductions and impressions revealed that more inaccessible traits were omitted than accessible traits. These two
studies indicate the viability of considering chronic differences in
construct accessibility, and the importance of such accessibility
for memory and impressions of people. Especially noteworthy is
the fact that such effects persisted over a two-week period.

Summary

In this section we reviewed and reformulated issues addressed
by the traditional language and thought literature in terms of construct accessibility. We also discussed the implications of such
issues for the determinants and consequences of communication
and language use. We view individual variation in construct ac-
cessibililty as often arising from differing linguistic and socialization backgrounds, and as having direct and indirect implications
for both information processing and several forms of communication activity. It is possible that differences in construct accessibility (both momentary and chronic) may underlie documented differences in language use (Nunally, 1965).

CONCLUSIONS

We have presented two approaches to individual differences in
language use—the communication game and construct
accessiblity—that we believe can benefit, as well as benefit from,
social psychological analyses of human behavior. Both of these
approaches are based on current work in social cognition.
Throughout this chapter we have attempted to integrate theory
and research from diverse areas in order to demonstrate the utility of these formulations. One value of these approaches is their
ability to unify diverse areas. Each approach has practical implications as well. The social class differences described earlier, for example, argue for a more systematic examination of individual dif-

ferences in communication goal choice and rule-following as a function of socialization. This would seem to be a prerequisite for attempts to provide adequate educational opportunities for individuals of different socioeconomic backgrounds (see also Higgins, Fondacaro, et al., 1981). We suggested that social psychology has paid little attention to individual differences in communication and language use. Such analyses are needed for several reasons. First, interpersonal communication is an important aspect of social behavior and, therefore, should receive attention from the social branch of psychology. Second, the importance of such individual differences has recently beome evident in several areas of interest to social psychologists—for example, the social psychology of law and educational psychology. Although in its initial stages, it is instructive to consider the relevance of communication issues to this research.

Lind and O'Barr (1979) and Scherer (1979) have examined the impact of individual differences in language use in the legal context. They point to the potential importance of such differences for both witness testimony and jury decision making. Lind and O'Barr examined the impact of language variables on reactions to hypothetical courtroom witnesses. One dimension, power versus powerless speech, has been shown to distinguish males from females. Females, for example, use frequent hedges, empty adjectives, and intensifiers. Lind and O'Barr have suggested that this type of speech is not only characteristic of women, but reflects a lower social status generally and is thus characteristic of powerless groups. To examine the impact of powerless speech in the courtroom, they played tapes of hypothetical testimony with subjects acting as jurors. While trial evidence remained constant, speech varied. Witnesses were then rated on a series of scales such as competence, trustworthiness, convincingness, and so on. As expected, witnesses who used powerless speech were rated more negatively than were the other witnesses. Scherer (1979) examined the implications of speech style for jurors and found that speech styles correlated with juror influence during deliberations, but that the particulars of an influential style varied crossculturally. From work such as this, it is clear that language and communication variables are important to any comprehensive anlaysis of the legal process.

Language use also has implication for education, particularly teachers' expectations of student performance (Rosenthal & Jacobsen, 1968). Edwards (1979), for example, examined the effect of social class speech differences on the reactions of teachers to their students. These middle-class teachers were asked to listen to the tape-recorded speech of lower-or middle-class children. They then rated the children on a variety of scales assessing such attributes as status, intelligence, and vocabulary. The results indicated that lower-class children were rated lower than middle-class children on all scales. To the extent that such speech distinctions set up teacher expectancies in the classroom, lower-class children may face a considerable educational barrier. A more complete understanding of the functions that such social class differences in speech play in various social groups is needed in order that comparable educational experiences may be provided for students from different speech communities.

The approaches presented in this chapter also have implications for current treatments of psychopathology. The possible value of the communication game approach is intimated by the role that communication has assumed in many analyses of psychopathology (for example, Mischler & Waxler, 1965; Reusch & Bateson, 1968; Riskin & Faunce, 1972). Consider, for example, Reusch and Bateson's "social matrix" analysis of mental illness. They have defined mental illness as a disturbance in communication. Neuroses, for example, are seen as attempted manipulations on the part of neurotics "to create a stage to convey messages to other more effectively" (1968, p.88). When these communications fail, the individual experiences new frustrations with which he must cope. Moreover, given that the therapeutic process in their view is a dynamic interaction of two communication systems (patient and therapist), the cognitive simlarity of the interactants, including the similarity of their accessible constructs, may be important to the success of the therapy (see Menges, 1968; Runkel, 1956; Triandis, 1960). The match between the communication goals of the client and the therapist (as well as the goals emphasized by the mode of therapy itself) may also influence the efficiency and effectiveness of the therapeutic process, which involves a variety of potentially conflicting goals (for example, "face" goals, social relationship goals, and task goals). Thus, knowledge concerning the determinants and consequences of individual differences in communicative orientation may enhance the success of the therapeutic process.

Finally, communication differences provide part of the answer to the classic question, "Why do different people judge and remember social events differently?" Is it because they actually perceive or process information differently? Perhaps. We would suggest, however, that at least some of these differences occur because individuals communicate differently about social events. These differences may arise from differences in communication goals and the accessibility of constructs, which may in turn cause differences in attitudes, judgments, and memory. From this perspective, the Sapir Whorf notion of linguistic determinism is alive and well, but at the level of personal style rather than the language system per se, and as a factor that is itself determined by social experience and motivation.

REFERENCES

Adorno, T. W., Frenkel-Brunswick, E., Levinson, D. J., & Sanford, R. N. (1950). *The authoritarian personality*. New York: Harper & Row.

Amir, Y. (1969). Contact hypotheses in ethnic relations. *Psychological Bulletin 71*, 319-342.

Austin, J.L. (1962). *How to do things with words*. Oxford: Oxford University Press.

Bales, R.F. (1955). The equilibrium problem in small groups. In A.P. Hare, E. F. Borgatta, & R.F. Bales (Eds.), *Small groups: Studies in social interaction*. New York: Knopf.

Barron, N. (1971). Sex-type language: The product of grammatical cases. *Acta Sociologica 14*, 24-72.

Bartlett, F.C. (1932) *Remembering*. Cambridge: University Press.

Beach, L., & Wertheimer, M. (1961). A free response apprach to the study of person cognition. *Journal of Abnormal and Social Psychology 62*, 367-374.

Berg, K.S., & Vidmar, N. (1975). Authoritarianism and recall of evidence about criminal behavior. *Journal of Research in Personality 9*, 147-157.

Berlin, B., & Kay, P. (1969). *Basic color terms: Their universality and evolution*. Berkeley: University of California Press.

Bermont, G., Nemeth, C. & Vidmar, N.J. (Eds.). *Psychology and the law*. Lexington, MA: Lexington.

Bernstein, B. (1960). Language and social class. *British Journal of Sociology 11*, 271-276.

Bernstein, J. (1975). *Another look at sex differentiation in the socialization behaviors of mothers and fathers*. Paper presented at the Conference for New Directions for Research on the Psychology of Women, Madison, Wisconsin.

Blumer, H. (1962). Society as symbolic interaction. IN A.M. Rose (Ed.), *Human behavior and social processes*. London: Routledge & Kegan Paul.

Bodine, A. (1975). Sex differentiation in language. In B. Thorne & N. Henley (Eds.), *Language & sex: Difference & dominance*. Rowley, MA: Newbury.

Bolinger, D. (1975). *Aspects of language* (2nd ed.). New York: Harcourt Brace Janovich.

Bourhis, R.U., & Giles, H. (1977). The language of intergroup distinctiveness. In H. Giles (Ed.) *Language, ethnicity and intergroup relations*. London: Academic Press.

Bourhis, R.U. Giles, H., Leyens, J. P., & Tajfel, H. (1979). Psycholinguistic distinctiveness: Language divergence in Belgium. In H. Giles & R.N. St. Clair (Eds.), *Language and social psychology*. Oxford: Blackwell.

Brown, R. W. (1965). *Social psychology*. New York: Free Press.

Brown, R.W., & Lennenberg, E. H. (1954). A study in language and cognition. *Journal of Abnormal and Social Psychology 49*, 454-462.

Bruner, J. S. (1957). On perceptual readiness. *Psychological Review, 64*, 123-152.

Bruner, J. S. (1958). Social psychology and perception. In E. E. Maccoby, T. M. Newcomb, & E. L. Hartley (Eds.), *Readings in social psychology*. New York: Holt, Rinehart & Winston.

Byrne, D. (1971). *The attraction paradigm*. New York: Academic Press.

Cantor, N., & Kihlstrom, J. F. (1981). *Personality, cognition, and social interaction*. Hillsdale, NJ: Lawrence Erlbaum.

Carlston, D. E. (1980). Event, inferences, and impression formation. In R. Hastie, T. M. Ostrom, E. B. Ebbesen, R. S. Wyer, Jr., D. L. Hamilton, D. E. Carlston (Eds.), *Person memory: The cognitive basis of social perception*. Hillsdale, NJ: Lawrence Erlbaum.

Carmichael, L., Hogan, H. P., & Walter, A. A. (1932). An experimental study of the effect of language on the reproduction of visually perceived form. *Journal of Experimental Psychology, 15*, 72-86.

Carroll, J. B., & Casagrande, J. B. (1958). The functions of language classification in behavior. In E. E. Maccoby, T. M. Newcomb, & E. L. Hartley (Eds.), *Readings in social psychology*. New York: Holt, Rinehart & Winston.

Cazden, C. B. (1968). Three sociolinguistic views of the language and speech of lower-class children: With special attention to the work of Basil Bernstein. *Developmental Medicine and Child Neurology, 19*, 600-612.

Cazden, C. B. (1970). The situation: A neglected source of social class differences in language use. *Journal of Social Issues, 26*, 35-60.

Cazden, C. B., Cox, M., Dickinson, D., Steinberg, Z., & Stone. C. (1979). "You all gonna hafta listen": Peer teaching in a primary classroom. In W. A. Collins (Ed.), *Children's language and communication*. Hillsdale, NJ: Lawrence Erlbaum, 1979.

Cazden, C. B., John, U. P., & Hymes, D. (1972). *Functions of language in the classroom*. New York: Columbia University Press.

Chapman, L. J., & Chapman, J. P. (1969). Illusory correlation as an obstacle to the use of valid psychodiagnostic signs. *Journal of Abnormal Psychology, 74*, 271-280.

Chomsky, N. (1965). *Aspects of the theory of syntax*. Cambridge, MA: MIT Press.

Chomsky, N. (1968). *Language and mind*. New York: Harcourt Brace Jovanovich.

Clark, R. A., & Delia, J. G. (1979). Topoi and rhetorical competence. *Quarterly Journal of Speech, 65*, 187-206.

Cohen, A. R. (1969). Cognitive tuning as a factor affecting impression formation. *Journal of Personality, 29*, 235-245.

Cohen, C. E. (1981). Goals and schemata in person perception: Making sense from the stream of behavior. In N. Cantor & J. F. Kihlstrom (Eds.) *Personality, cognition, and social interaction*. Hillsdale, NJ: Lawrence Erlbaum.

Cohen, C. E., & Ebbesen, E. B. (1979). Observational goals and schema activation: A theoretical framework for behavior perception. *Journal of Experimental Social Psychology, 15*, 305-329.

Conklin, W. F. (1974). Toward a feminist analysis of linguistic behavior. *University of Michigan Papers in Women's Studies, 1*, 51-73.

Cushman, D., & Whiting, G. C. (1972). An approach to communication theory: Toward consensus on rules. *Journal of Communication, 22*, 217-238.

Delia, J. G. (1976). A constructionist analysis of the concept of credibility. *Quarterly Journal of Speech, 62,* 361-375.

Deutsch, M. (1962). Cooperation and trust: Some theoretical notes. In M. R. Jones (Ed.,) *Nebraska symposium on motivation.* Lincoln: University of Nebraska Press.

Dornbusch, S., Hastorf, A., Richardson, S. A., Muzzy, R., & Vreeland, R. S. (1965). The perceiver and the perceived: Their relative influence on the categories of interpersonal perception. *Journal of Personality and Social Psychology, 1,* 434-440.

Dunn, S., Bliss, J., & Siipilo, K. (1958). Effects of impulsivity, introversion, and individual values upon association under free conditions. *Journal of Personality, 26,* 61-76.

Eagly, A. H., & Himmelfarb, S. (1978). Attitudes and opinions. *Annual Review of Psychology, 29.*

Eagly, A. H., Wood, W., & Chaiken, S. (1978). Causal inferences about communicators and their effect on opinion change. *Journal of Personality and Social Psychology, 36,* 424-435.

Eakins, B. W., & Eakins, R. G. (1978). *Sex differences in human communication.* Boston: Houghton Mifflin.

Edwards, J. R. (1979). Judgments and confidence in reactions to disadvantaged speech. In H. Giles & R. N. St. Clair (Eds.), *Language and social psychology.* Oxford: Blackwell.

Ervin-Tripp, s. (1967). Sociolinguistics. In L. Berkowitz (Ed.), *Advances in experimental social psychology: Vol. 4.* New York: Academic Press.

Festinger, L. (1950). Informal social communication. *Psychological Review, 57,* 271-282.

Festinger, L. (1954). A theory of social comparison processes. *Human Relation, 7,* 117-140.

Fiedler, F. (1971). *Leadership.* New York: General Learning Press.

Fishman, J. A. (1960). A systematization of the Whorfian hypothesis. *Behavioral Science, 5,* 323-339.

Flavell, J. H., Botkin, P. T., Fry, C. L., Wright, J. W., & Jarvis, P. E. (1968). *The development of role-taking and communication skills in children.* New York: John Wiley.

Foley, J. P., & Macmillan, Z. L. (1943). Mediated generalization and the interpretation of verbal behavior: V. "Free association" as related to differences in professional training. *Journal of Experimental Psychology, 33,* 299-310.

Forbach, G. B., Stanners, R. F., & Hochhaus, L. (1974). Repetition and practice effects in a lexical decision task. *Memory and Cognition, 2,* 337-339.

French, J. R., & Raven, B. (1959). The bases of social power. In D. Cartwright (Ed.), *Studies in social power.* Ann Arbor, MIj: Institute of Social Relations.

Garfinkel, H. (1967). *Studies in ethnomethodology.* Englewood Cliffs, NJ: Prentice-Hall.

Garvey, C., & Hogan, R. (1973). Social speech and social interaction: Egocentrism revisited. *Child Development, 44,* 562-568.

Giles, H., & Smith, P. M. (1979). Accommodation theory: Optimal levels of convergence. In H. Giles & R. St. Clair (Eds.), *Language and social psychology.* Oxford: Blackwell.

Glanzer, M., & Clark, W. H. (1962). Accuracy of perceptual recall: An analysis of organization. *Journal of Verbal Learning and Verbal Behavior, 1,* 225-242.

Gleser, G. D., Gottschalk, L., & Watkins, J. (1959). The relationship of sex and intelligence to choice of words: A normative study of verbs and behavior. *Journal of Clinical Psychology, 15,* 182-191.

Glucksberg, S., Krauss, R. M., & Higgins, E. T. (1975). The development of referential communication skills. In F. Horowitz, E. Hetherington, S. Scarr-Salapetek, & G. Siegel (Eds.), *Review of child development research: Vol. 4.* Chicago: University of Chicago Press.

Glucksberg, S., & Weisberg, R. W. (1966). Verbal behavior and problem solving: Some effects of labeling in a functional fixedness problem. *Journal of Experimental Psychology, 71,* 659-664.

Goffman, E. (1959). *The presentation of self in everyday life.* Garden City, NY: Doubleday.

Goffman, E. (1967). *Interaction ritual: Essays on face-to-face behavior.* Garden City, NY: Doubleday.

Grice, H. P. (1971). Logic and conversation: The William James lectures, Harvard University, 1967-68. In P. Cole & J. L. Morgan (Eds.) *Syntax and semantics: Vol. 3. Speech acts.* New York: Academic Press.

Gumperz, J. J., & Hymes, D. (Eds.). (1972). *Directions in sociolinguistics: The ethnography of communication.* New York: Holt, Rinehart & Winston.

Hamilton, D. L. (Ed.). (1981). *Cognitive processes in stereotyping and intergroup behavior.* Hillsdale, NJ: Lawrence Erlbaum.

Hamilton, D. L., & Gifford, R. K. (1976). Illusory correlation in interpersonal perception: A cognitive basis of stereotypic judgment. *Journal of Experimental Social Psychology, 13,* 392-407.

Hamilton, D. L., Katz, L., & Leirer, U. (1980). Organizational processes in impression formation. In R. Hastie, T. M. Ostrom, D. L. Hamilton, E. Ebbesen, R. S. Wyer, & D. Carlson (Eds.), *Person memory: The cognitive basis of social perception.* Hillsdale, NJ: Lawrence Erlbaum.

Harkins, S. G., Harvey, J. H., Keithly, L. & Rich, M. (1977). Cognitive tuning, encoding and the attribution of causality. *Memory and Cognition, 5,* 561-565.

Harvey, O. J., & Beverly, G. D. (1961). Some personality correlates of concept change through role playing. Journal of Abnormal and Social Psychology, 63, 125-130.

Hastie, R., Ostrom, T. M., Hamilton, D. L., Ebbesen, E., WWyer, R. S., & Carlston, D. (Eds.). (1980). *Person memory: The cognitive basis of social perception.* Hillsdale, NJ: Lawrence Erlbaum.

Hawes, L. C. (1973). Elements of a model for communication processes. *Quarterly Journal of Speech, 59,* j11-21.

Hawkins, P. R. (1969). Social class, the nominal group and reference. *Language and Speech, 12,* 125-135.

Heider, E. R. (1971). Style and accuracy of verbal communications within and between social classes. *Journal of Personality and Social Psychology, 18,* 33-47.

Heider, E. R. (1972). Universals in color naming and memory. *Journal of Experimental Psychology, 93,* 10-20.

Higgins, E. T. (1976). Social class differences in verbal communicative accuracy: A question of "which question?" *Psychological Bulletin, 83,* 695-674.

Higgins, E. T. (1977). Communication development as related to channel, incentive, and social class. *Genetic Psychology Monographs, 96,* 75-141.

Higgins, E. T. (1981). The "communication game": Implications for social cognition and persuasion. In E. T. Higgins, C. P. Herman, & M. P. Zanna (Eds.), *Social cognition: The Ontario symposium.* Hillsdale, NJ: Lawrence Erlbaum.

Higgins, E. T., Fondacaro, R., & McCann, C. D. (1981). Rules and roles: The "communication game" and speaker-listener processes. In W. P. Dickson (Ed.), Children's oral communication skills. New York: Academic Press.

Higgins, E. T., Herman, C. P., & Zanna, M. P. (Eds.). (1981). Social cognition: The Ontario symposium. Hillsdale, NJ: Lawrence Erlbaum.

Higgins, E. T., & King, G. A. (1981). Accessibility of social constructs: Information processing consequences of individual and contextual variability. In N. Cantor & J. H. Kihlstrom (Eds.), Personality, cognition, and social interaction. Hillsdale, NJ: Lawrence Erlbaum.

Higgins, E. T., King, G. A., & Mavin, G. H. (1982). Individual construct accessibility and subjective impressions and recall. Journal of Personality and Social Psychology, 43, 35-47.

Higgins, E. T., & Krauss, R. M. (1973). Grammatical categorization and functional fixedness: A test of linguistic determinism. New Orleans: American Psychological Association.

Higgins, E. T., McCann, C. D., & Fondacaro, R. (1982). The "communication-game": Goal-directed encoding and cognitive consequences. Social Cognition, 1, 21-37.

Higgins, E. T., & Rholes, W. S. (1978). "Saying is believing": Effects of message modification on memory and liking for the person described. Journal of Experimental Social Psychology, 14, 363-378.

Higgins, E. T., Rholes, W. S., & Jones, C. R. (1977). Category accessibility and impression formation. Journal of Experimental Social Psychology, 13, 141-154.

Huttenlocher, J., & Higgins, E. T. (1971). Adjectives, comparatives, and syllogisms. Psychological Review, 78, 487-504.

Hymes, D. (1979). Sapir, competence, voices. In C. J. Fillmore, D. Kempler, & W. S-Y. Wang (Eds.), Individual differences in language ability and language behavior. New York: Academic Press.

Isen, A. M., Shalken, T. L., Clark, M., & Karp, L. (1978). Affect, accessibility of material in memory, and behavior: A cognitive loop? Journal of Personality and Social Psychology, 36, 1-12.

Jeffery, K. M., & Mischel, W. (1979). Effects of purpose on the organization and recall of information in person perception. Journal of Personality, 47, 397-419.

Jenkins, J. J. (1960). Commonality and associations as an indication of more general patterns of verbal behavior. In T. A. Sebeok (Ed.), Style in language. New York: John Wiley.

Johnson, R. P. (1977). Social class and grammatical development: A comparison of the speech of five year olds from middle and working class backgrounds. Language and Speech, 20, 317.

Johsnon, R. P., & Singleton, C. H. (1977). Social class and communication style: The ability of middle and working class five year olds to encode and decode abstract stimuli, British Journal of Psychology, 68, 237-244.

Jones, E. E. (1964). Ingratiation: A social psychological analysis. New York: Appleton-Century-Crofts.

Jones, E. E., & Thibaut, J. W. (1958). Interaction goals as bases of inference in interpersonal perception. In R. Tagiuri & L. Petrullo (Eds.), Person perception and interpersonal behavior. Stanford, CA: Stanford University Press.

Kelley, H. H. (1979). Personal relationships: Their structures and processes. Hillsdale, NJ: Lawrence Erlbaum.

Kelley, H. H., & Stahelski, A. J. (1970). The social interaction basis of cooperators' and competitors' beliefs about others. *Journal of Personality and Social Psychology, 16,* 66-91.

Kelly, G. A. (1955). *The psychology of personal constructs.* New York: W. W. Norton.

Key, M. R. (1975). *Male/female language. Metuchen, NJ: Scarecrow.*

Krauss, R. M., Vivekanathan, P. S., & Weinheimer, X.X. (1968). "Inner speech" and "external speech": Characteristics and communicative effectiveness of socially and non-socially encoded messages. *Journal of Personality and Social Psychology, 9,* 295-300.

Labov, W. (1970). The logic of non-standard English. In F. Williams (Ed.), *Language and poverty: Perspectives on a theme.* Chicago: Markham.

Lakoff, R. (1973). Language and woman's place. *Language in Society, 2,* 45-79.

Lakoff, R. (1975). *Language and woman's place.* New York: Harper & Rowe.

Lakoff, R. T. (1979). Stylistic strategies within a grammar of style. In J. Orasanu, M. K. Slater, & L. L. Adler, (Eds.), Language, sex and gender: Does la difference make a difference? *Annals of the New York Academy of Science, 327,* 53-80.

Lambert, W. E., Hodgson, R. C., Gardner, R. C., & Fillenbaum, S. (1960). Evaluative reactions to spoken languages. *Journal of Abnormal and Social Psychology, 60,* 44-51.

Lantz, D., & Stefflre, V. (1964). Language and cognition revisited. *Journal of Abnormal and Social Psychology, 69,* 471-481.

Lenneberg, E. H. (1971). Language and cognition. In D. D. Steinberg & L. a. Jokobovits (Eds.), *Semantics: An interdisciplinary reader in philosophy, linguistics and psychology.* Cambridge: Cambridge University Press.

Leventhal, H. (1962). The effects of set and discrepancy on impression change. *Journal of Personality, 20,* 1-15.

Levine, R., Chein, I., & Murphy, G. (1942). The relation of the intensity of a need to the amount of perceptual distortion: A preliminary report. *Journal of Psychology, 13,* 283-293.

Lewis, H. B. (1944). An experimental study of the role of the ego in work: I. The role of the ego in cooperative work. *Journal of Experimental Psychology, 34,* 113-127.

Lewis, H. B., & Franklin, M. (1944). An experimental study of the role of the ego in work: II. The significance of task orientation in work. *Journal of Experimental Psychology, 34,* 194-215.

Lind, E. A., & O'Barr, W. M. (1979). The social significance of speech in the courtroom. In H. Giles and R. St. Clair (Eds.), *Language and social psychology.* Oxford: Blackwell.

Lingle, J. H., & Ostrom, T. M. (1979). Retrieval selectivity in memory-based judgments. *Journal of Personality and Social Psychology, 37,* 180-194.

Loftus, E. F., & Palmer, J. C. (1974). Reconstruction of automobile destruction: An example of the interaction between language and memory. *Journal of Verbal Learning and Verbal Behavior, 13,* 585-589.

Lucy, J. A., & Shweder, R. A. (1981). Whorf and his critics: Linguistic and nonlinguistic influence on color memory. In R. W. Casson, (Ed.), *Language, culture, and cognition: Anthropological perspectives.* New York: Macmillan.

Lyman, S. M., & Scott, M. B. (1970). *A sociology of the absurd.* New York: Appleton-Century-Crofts.

Malpass, R. S., & Devine, P. G. (1981a). Guided memory in eyewitness identification. *Journal of Applied Psychology, 66,* 343-350.

Malpass, R. S., & Devine, P. G. (1981b). Eyewitness identification: Lineup instructions and the absence of the offender. *Journal of Applied Psychology, 66,* 482-489.

Mandelbaum, D. G. (Ed.). (1947). *Selected writings of Edward Sapir.* Berkeley: University of California Press.

Manis, M., Cornell, S. D., & Moore, J. C. (1974). Transmission of attitude-relevant information through a communication chain. *Journal of Personality and Social Psychology, 30,* 81-94.

Markus, H. (1977). Self-schemata and processing information about the self. *Journal of Personality and Social Psychology, 35,* 63-78.

McCann, D. (1982). *Individual differences in social communication: Self-monitoring and the "communication game."* Paper presented at the 90th Annual Convention of the American Psychological Association, August.

McCann, D., & Hancock, R. D. (in press). Self-monitoring in communicative interactions: Social-cognitive consequences of goal-directed message modification. *Journal of Experimental Social Psychology.*

McConnell-Ginet, S. (1979). Our father tongue: Essays in linguistic politics. *Diacritics,* 44-50.

McGuire, W. J. (1969). The nature of attitudes and attitude change. In G. Lindzey & E. Aronson (Eds.), *The handbook of social psychology.* Reading, MA: Addison-Wesley.

McGuire, W. J. (1972). Attitude change: The information-processing paradigm. In C. G. McClintock (Ed.), *Experimental social psychology.* New York: Holt, Rinehart & Winston.

Mead, G. H. (1934). *Mind, self, and society.* Chicago: University of Chicago Press.

Mehrabian, A., & Reed, H. (1968). Some determinants of communication accuracy. *Psychological Bulletin, 70,* 365-381.

Menges, R. J. (1968). Student-instructor cognitive compatibility in the large lecture class. *Journal of Personality, 37,* 444-459.

Mills, J., & Jellison, J. M. (1967). Effect on opinion change of how desirable the communication is to the audience the communicator addressed. *Journal of Personality and Social Psychology, 6,* 98-101.

Mischler, E. G., & Waxler, W. E. (1965). Family interaction process and schizophrenia: A review of current theories. *Merrill-Palmer Quarterly of Behavior and Development, 11,* 269-315.

Moreland, R. L., & Levine, J. M. (in press). Socialization in small groups: Temporal changes in individual-group relations. In L. Berkowitz (Ed.), *Advances in experimental social psychology: Vol. 15.* New York: Academic Press.

Morris, C. (1964). *Signification and significance.* Cambridge, MA: MIT Press.

Moscovici, S. (1967). Communication processes and the properties of language. In L. Berkowitz (Ed.), *Advances in experimental social psychology: Vol. 3.* New York: Academic Press.

Neisser, U. (1967). *Cognitive psychology.* New York: Appleton-Century-Crofts.

Neisser, U. (1976). *Cognition and reality: Principles and implications of cognitive psychology.* San Francisco: Freeman.

Newtson, D., & Czerlinsky, T. (1974). Adjustment of attitude communications for contrasts by extreme audiences. *Journal of Personality and Social Psychology, 30,* 829-837.

Nisbett, R. E., & Ross, L. D. (1980). *Human inference: Strategies and shortcomings of informal judgment.* Englewood Cliffs, NJ: Prentice-Hall.

Nunally, J. C. (1965). Individual differences in word usage. In S. Rosenberg (Ed.), *Directions in Psycholinguistics.* New York: Macmillan.

O'Keefe, B. J., & Delia, J. G., & O'Keefe, D. J. (1977). Construct individuality, cognitive complexity, and the formation and remembering of interpersonal impressions. *Social Behavior and Personality, 5,* 229-240.

Padgett, V. R., & Wolosin, R. J. (1980). Cognitive similarity in dyadic communication. *Journal of Personality and Social Psychology, 39,* 654-659.

Peirce, C. S. (1940). Logic as semiotic: The theory of signs. In J. Buchler (Ed.), *The philosophy of Peirce: Selected writings.* London: Routledge & Kegan Paul.

Philips, S. U. (1980). Sex differences and language. *Annual Review of Anthropology: Vol. 9.* Annual Review.

Piaget, J. (1926). *The language and thought of the child.* New York: Harcourt Brace Jovanovich.

Posner, M. I. (1978). *Chronometric explorations of the mind.* Hillsdale, NJ: Lawrence Erlbaum.

Postman, L., & Brown, D. R. (1952). The perceptual consequences of success and failure. *Journal of Abnormal and Social Psychology, 47,* 213-221.

Ranken, H. B. (1963). Language and thinking: Positive and negative effects of naming. *Science, 141,* 48-50.

Reed, S. K. (1972). Pattern recognition and categorization. *Cognitive Psychology, 3,* 382-407.

Reusch, J., & Bateson, G. (1968). *Communication: The social matrix of psychiatry.* New York: W. W. Norton.

Riskin, J., & Faunce, E. E. (1972). An evaluative review of family interaction research. *Family Process, 11,* 365-455.

Rommetveit, R. (1974). *On message structure: A framework for the study of language and communication.* New York: John Wiley.

Rosch, E. (1978). Principles of categorization. In. E. Rosch & B. B. Lloyd (Eds.), *Cognition and categorization.* Hillsdale, NJ: Lawrence Erlbaum.

Rosch, E., Mervis, C. B. (1975). Family resemblances: Studies in the internal structure of categories. *Cognitive Psychology, 7,* 573-605.

Rosenthal, R., & Jacobson, L. (1968). *Pygmalion in the classroom: Teacher expectancies and pupils' intellectual development.* New York: Holt, Rinehart & Winston.

Ross, L., Lepper, M. R., Strack, R., & Steinmetz, J. (1977). Social explanation and social expectation: Effects of real and hypothetical explanations on subjective likelihood. *Journal of Personality and Social Psychology, 35,* 817-829.

Ross, M., & Sicoly, F. (1979). Egocentric biases in availability and attribution. *Journal of Personality and Social Psychology, 37,* 322-336.

Rothbart, M., Fulero, S., Jensen, C., Howard, J., & Birrell, P. (1978). From individual to group impressions: Availability heuristics in stereotype formation. *Journal of Experimental Social Psychology, 14,* 235-255.

Runkel, P. (1956). Cognitive similarity in facilitating communication. *Sociometry, 19,* 178-191.

Ryan, E. B. (1979). Why do low-prestige language varieties persist? In H. Giles & R. N. St. Clair (Eds.), *Language and social psychology.* Oxford: Blackwell.

Scribner, S., & Cole, M. (in press). Literacy without schooling: Testing for intellectual effects. *Harvard Educational Review.*

Searle, J. R. (1976). A classification of illocutionary acts. *Language in Society, 5,* 1-23.

Shaw, M. E., & Costanzo, P. R. (1970). *Theories of social psychology.* New York: McGraw-Hill.

Scherer, K. R. (1979). Voice and speech correlates of perceived social influence in simulated juries. In H. Giles & R. W. St. Clair (Eds.), *Language and social psychology.* Oxford: Blackwell.

Sherman, S. J., Ahlm, K., Berman, L., & Lynn, s. (1978). Contrast effects and their relationship to subsequent behavior. *Journal of Experimental Social Psychology, 14,* 340-350.

Sherif, M., Harvey, O., White, B., Hood, W., & Sherif, C. (1961). *Intergroup conflict and cooperation: The robber's cave experiment.* Norman: University of Oklahoma Press.

Smith, E. E., Shoben, E. J., & Rips, L. J. (1974). Structure and process in semantic memory: A feature model for semantic decisions. *Psychological Review, 81,* 214-241.

Snyder, M. (1974). Self-monitoring of expressive behavior. *Journal of Personality and Social Psychology, 30,* 526-537.

Snyder, M. (1979). Self-monitoring processes. In L. Berkowitz (Ed.), *Advances in experimental social psychology: Vol. 12.* New York: Academic Press.

Spiro, R. J. (1977). Remembering information from test: The state of the "schema" approach. In R. C. Anderson, R. J. Spiro, & W. E. Montague (Eds.), *Schooling and the acquisition of knowledge.* Hillsdale, NJ: Lawrence Erlbaum.

Srull, t. K., & Wyer, R. S. (1979). The role of category accessibility in the interpretation of information about persons: Some determinants and implications. *Journal of Personality and Social Psychology, 37,* 1660-1672.

Stefflre, V., Vales, V. C., & Morley, L. (1966). Language and cognition in Yucatan: A cross cultural replication. *Journal of Personality and Social Psychology, 4,* 112-115.

Straus, M. A. (1968). Communication, creativity, and problem solving ability of middle and working-class families in three societies. *American Journal of Sociology, 73,* 417-430.

Strauss, A., & Schatzman, L. (1955). Cross-class interviewing: An analysis of interaction and communication styles. *Human Organization, 14,* 28-31.

Tajfel, H. (1974). Social identity and intergroup behavior. *Social Science Information, 13,* 65-93.

Tajfel, H. (Ed.). (1979). Differentiation between social groups: Studies in the social psychology of intergroup behavior. *European monographs in social psychology: Vo. 14.* London: Academic Press.

Tannen, D. (1981a). New York Jewish conversational style. *International Journal of the Society of Language, 30,* 133-149.

Tannen, D. (1981b). The machine-gun question: An example of conversational style. *Journal of Pragmatics, 30.*

Taylor, D. M., Bassili, J. N., & Aboud, F. E. (1973). Dimensions of ethnic identity: An example from Quebec. *Journal of Social Psychology, 89,* 185-192.

Taylor, D. M., Simard, L. M., & Aboud, F. E. (1972). Ethnic identification in Canada: A cross-cultural investigation. *Canadian Journal of Behavioral Science, 4,* 13-24.

Taylor, S. E., & Crocker, J. (1981). Schematic bases of information processing. In E. T. Higgins, C. P. Herman, & M. P. Zanna (Eds.), *Social cognition: The Ontario sympposium: Vol. 1.* Hillsdale, NJ: Lawrence Erlbaum.

Thakerar, J. N., Giles, H., & Cheshire, J. (in press). Psychological and linguistic parameters of speech accommodation theory. In C. Fraser & K. R. Sherer (Eds.), *Social psychological dimensions of language behavior.* Cambridge: Cambridge University Press.

Thibaut, J. W., & Riecken, H. W. (1955). Authoritarianism, status and the communication of aggression. *Human Relations, 8,* 95-120.

Thomas, d. R., & De Capito, A. (1966). Role of stimulus labeling in stimulus generalization. *Journal of Experimental Psychology, 71,* 913-915.

Thorne, B., & Henley, N. (Eds.). (1975). *Language and sex: Difference and dominance.* Rowley, MA: Newbury.

Triandis, H. C. (1960). Cognitive similarity and communication in a dyad. *Human Relations, 13,* 175-183.

Trudgill, P. (1975). *Sociolinguistics: An introduction.* Middlesex: Penguin.

Tubbs, S. L., & Moss, S. (1977). *Human communication* (2nd ed.). New York: Random House.

Turner, C. J., & Pickvance, R. E. (1971). Social class differences in the expression of uncertainty in five year old children. *Language and Speech, 14,* 303-325.

Tversky, A., & Kahneman, D. (1973). Availability: A heuristic for judging frequency and probability. *Cognitive Psychology, 5,* 207-232.

van Dijk, T. A. (1977). Context and cognition: Knowledge frames and speech act comprehension. *Journal of Pragmatics, 1,* 211-232.

Vygotsky, L. S. (1962). *Thought and language.* E. Hanfmann & G. Vakar (trans.). Cambridge, MA: MIT Press.

Walster, E., & Festinger, L. (1962). The effectiveness of "overheard" persuasive communications. *Journal of Abnormal and Social Psychology, 65,* 395-402.

Warren, N. (1966). Social class and construct systems: An examination of the cognitive structure of two social class groups. *British Journal of Social and Clinical Psychology, 5,* 254-263.

Warren, R. E. (1972). Stimulus encoding and memory. *Journal of Experimental Psychology, 94,* 90-100.

Watzlawick, P., Beavin, J. H., & Jackson, D. D. (1967). *Pragmatics of human communication.* New York; W. W. Norton.

Werner, H., & Kaplan, B. (1963). *Symbol formation.* New York: John Wiley.

Williams, G. (1979). Language group allegiance and ethnic interaction. In H. Giles & B. St. Jacques (Eds.), *Language and ethnic relations.* Oxford: Pergamon Press.

Wirtenberg, T., & Nakamura, C. (1976). Education: Barrier or boon to changing occupational roles of women. *Journal of Social Issues, 32,* 165-179.

Wispe, L. G., & Drambarean, N. C. (1953). Physiological need, word frequency, and visual duration thresholds. *Journal of Experimental Psychology, 46,* 25-31.

Wittgenstein, L. (1953). *Philosophical investigations.* New York: Macmillan.

Wolosin, R. J. (1975). Cognitive similarity and group laughter. *Journal of Personality and Social Psychology, 32,* 503-509.

Wood, M. M (1966). The influence of sex and knowledge of communication ieffectiveness on spontaneous speech. *Word, 22,* 12-127.

Wyer, R. S., & Srull, T. K. (1980). The processing of social stimulus information: A conceptual integration. In R. Hastie, E. B. Ebbesen, T. M. Ostrom, R. S. Wyer, D. L. Hamilton, & D. E. Carlston, (Eds.), *Person memory: The cognitive basis of social perception.* Hillsdale, NJ: Lawrence Erlbaum.

Yarrow, M. R., & Campbell, J. D. (1963). Person perception in children. *Merrill-Palmer Quarterly, 9,* 57-72.

Yarmey, A. D. (1974). *The psychology of eyewitness testimony.* New York: Free Press.

Zajonc, R. B. (1960). The process of cognitive tuning and communication. *Journal of Abnormal and Social Psychology, 61,* 159-167.

Zajonc, R. B. (1980). Feeling and thinking: Preferences need no inferences. *American Psychologist, 35,* 151-175.

8

Analyzing Conversations:
Context, Content, Inference, and Structure

Thomas J. Housel

The following is a conversation between Bill and Mary, who have been married for a number of years:

Bill: "Saw John today."

Mary: "Yea."

Bill: "He's getting a divorce."

Mary: "That explains. . .and why she. . ."

Bill: "That's right."

Mary: "Of course."

How is it that these two interactants understood each other perfectly, while outsiders like ourselves have some difficulty understanding this conversation? The broad question of how interactants make sense of one another's utterances has been the central issue in conversational research for some time. Communication theorists, ethnomethodologists, psychologists, linguists, and even artificial intelligence theorists have developed their own unique and insightful approaches to help us understand how interactants make and derive meaning from conversations.[1] While these scholars employ a number of distinct philosophical and theoretical positions, their work offers potentially useful insights on the study of conversational coherence. Regardless of the conversational analyst's perspective, however, certain central components must be addressed if we are to understand and analyze how interactants make sensible comments and make sense of each other's conversational inputs.

Readers will agree on those concepts universal to all conversations: All conversations occur within a social context; all conversations carry a variety of levels of meaning; all conversations have some form or structure; and all conversations require interactants to make inferences. Some of the steps conversational analysts must take to understand the role of context, contents, levels of meaning, structure, and inferences in describing how interactants engage in the conversational process will be pointed out in an attempt to develop a rationale for studying conversational comprehension. These concepts overlap, rather than being independent of one another, because the processes they represent occur simultaneously in conversations.

SOCIAL CONTEXT

While this writer recognizes that definitions of social context depend on the researcher's theoretical perspective, the following definition is offered as a point of departure for discussion of the role of social context in helping interactants understand each other. Generally speaking, social context comprises the interactants, the cultural norms influencing them, the history of their prior interactions, the topic of their conversation, the state of their world knowledge, their physical surroundings, their purposes for interacting, their facility with the language being spoken, and the point in time when the interaction takes place.

In the sample conversation introducing this chapter, very little information was given about its social context, but the reader probably supplied a context (Bransford & Johnson, 1972). It was probably assumed that Bill and Mary were members of the Caucasian, English-speaking culture. One may have guessed that they were fairly intimate or at least knew each other rather well. Based on the reader's knowledge of marriage relations, it was probably evident from the comment about John's getting a divorce that the speakers were focusing on John's marital problem or its resolution. The "she" Mary referred to was very likely assumed to be John's wife. At least some of these assumptions about the conversation's context surely occurred to the reader, even if not all of them were included.

Before analysis, a conversational researcher faced with this conversation would likely want to know more about its social context. The researcher might inquire as to the background the interactants: What was their relationship? Where and when did they have the discussion? Had they talked about this topic previously? What preceded and followed this discussion?

Some analysts would be satisfied to study primarily those externally observable aspects of the conversation's social context which did not require knowledge of the interactants' mental states (Beach, 1983; Hawes, 1977; Sacks, Schegloff, & Jefferson, 1974; Schegloff & Sacks, 1973; Schegloff, 1968). They would argue that because we cannot directly observe the interactants' mental states (intentions, motives, memory, and so on), we would be forced to rely on the interactants' reports of what their mental states were at the time they were interacting. These researchers also would want to avoid inferring what the interactants were thinking during the interaction, because the inferences might represent potentially inaccurate observations. Representatives of this position usually require that everything needed to describe social context be present in the external environment, or, in other words, that it be "public" (O'Keefe, 1979, p. 209).[2]

While this explanation is somewhat oversimplified, the primary purpose in referring to this approach to analyzing social contexts is to show that these analysts chose to deal with social context at a rather high level of abstraction. They focus on public or external patterns of conversational behavior that typify and give coherence to a conversant's social world. Individual conversations are interesting to these analysts only because they provide insight into interaction patterns typical in the interactant's social world.

Simply put, conversational analysts who choose to focus on the external environment of social context avoid the possible pitfall of erroneous reports by interactants about their mental states, and also the possibility of making inaccurate inferences about their mental states. In so doing, however, they are forced to a rather high level of abstraction because their descriptions of social context are not limited to a particular couple interacting as individuals, but rather as any two members of a given culture.

Another group of conversational analysts (for example, Cicourel, 1980; Goss, 1982; Housel, 1982; O'Keefe, Delia, & O'Keefe, 1980; Planalp, 1983) would want to know as much about the participants' mental states as possible in attempting to define the social context of this conversation (O'Keefe, 1979). They would argue that the social context should include the interactant's memory for prior interactions, motives for bringing up and commenting on the topic, and meanings ascribed to the contents of the conversation in terms of its affective or connotative meaning and nonrelational or denotative meaning. To collect this information about social context, these analysts would ask the interactants what they were thinking. While such an approach carries with it the inherent risks of secondhand information, it also has the potential for giving the researcher a more detailed picture of the social context for specific conversations, allowing the analyst to focus on the actual contents of the conversation and their meaning for the interactants. Both approaches to defining the social context within which this couple converse provide the analyst with insights into the role social context plays in the meaning-making process at different levels of analysis. This point has been articulated recently by Gumperz (1982), who contends that language can be analyzed at a variety of levels, each of which adds to our understanding of the meaning-making process.

Unfortunately, the notion of social context has proven to be a rather ambiguous variable in the communication literature, primarily because there are few standard definitions for context. However, one way to talk about social context is to examine topics of conversations. If we know what interactants are talking about (the contents of their conversation), we will come closer to having the same context they have for making sense of each other's comments. The next part of this essay investigates the role of content in conversational analysis.

CONVERSATIONAL CONTENT

The contents of a conversation refer to "what the interactants are talking about"—the topics they are discussing, and information about those topics (Fillmore, 1977, p. 79). Making a commitment to study content requires that the analyst answer several preliminary questions. Where do conversants get the contents of their conversations? How do conversants make sense or assign meaning to the contents of their conversations? At what level should we analyze contents (individual, dyadic, social, or cultural)? Are these contents organized in any identifiable way?

To answer the first and last questions, the analyst might look to a dictionary that represents our current, culturally acceptable knowledge, but the dictionary says nothing about how interactants might combine such knowledge in unique ways to create new meanings. For those analysts wanting to provide more detailed analysis of how interactants use knowledge, the dictionary offers an undesirably high level of abstraction. For them, conversation content comes from interactants and is organized in ways that make it easy for interactants to use their knowledge to create meanings. To get the contents for their conversations, interactants use cognitive processes—in particular, memory— to retrieve and update their world knowledge. Perhaps, as early Chomskian linguistic theory proposed (1965), interactants have mental dictionaries from which they draw.

Why make any attempt to look into interactants mental dictionaries if we have access to current dictionaries which represent the potential contents contained in any individual interactant's mental dictionary? While this position appeals to some conversational analysts, because it avoids the sticky problems of where interactants get the contents of their conversation, it is severely limiting. In order to retain valuable contents of conversation for future use, interactants must store those contents in memory, but as recent research on memory points out (Clark & Clark, 1977; Dodd & White, 1981; Lachman, Lachman, & Butterfield, 1979), the notion of memory organized as a dictionary is untenable.

In conceptualizing memory, the notion of "schema" in modern information processing literature appears to provide a useful metaphor for memory storage, updating, and retrieval processes. Work on schemata or "schema theory," as it has come to be called, attempts to answer questions about where conversants get contents, how they make sense and assign meaning to contents, and how contents are organized. It also deals with contents at a variety of levels of abstraction—from the situation-specific contents of an ongoing conversation (Gumperz, 1982; Housel, 1982), to the cultural level of contents (Rice, 1980).

In addressing the content issue, Dillon (1981) uses schema theory to nake a distinction between what he calls content and structural schema. Dillon argues that texts or conversations can be characterized by content schemata which interactants are assumed to access and use to interpret the contents of a conversation. For example, when we discuss playing golf, we activate our golf content schema and "interpret a particular sequence of facts and details as an instance of playing golf" (Dillon, 1981, p. 51). Dillon further argues that this allows us to fill in "many particulars and shapes understanding of individual words, as in this case, for example *the ball*" (1981, p. 52). (Structural schema will be discussed in the "structure" section.)

Schema theory assumes that interactants store and retrieve conversation content in a hierarchically organized memory. The contents of memory are organized into schema which link related concepts. Thus, when Bill mentions John to Mary, Mary activates and enters her person schema (Owens, Bower, & Black, 1979; Cantor & Mischel, 1979; Hastie, Ostrom, Ebbesen, Wyer, Hamilton, & Carlston, 1980) for John. She uses this "John schema" to anticipate and help her make sense of the content of Bill's comments about John. Since Mary knows John, she can enter this specific schema. Had Mary not known John, she would have been forced to a higher level of abstraction and would have retrieved her schema for men in her culture, or perhaps a bit less abstract schema for Bill's male friends.

As Bill tacitly asks Mary to call up her schema for John, Mary anticipates that the contents of Bill's next statements will reinforce, change, or update her knowledge of John (see Hastie et al., 1980, for a comprehensive discussion of this phenomena). Bill may have multiple purposes (O'Keefe & Delia, 1982) for bringing up John. He may want Mary's opinion about the new content regarding John, as well as updating Mary's information about John, or he may simply want to converse with Mary in order to establish or maintain rapport (Lakoff, 1982).

Schema theory also postulates motive or purpose schemata (Wyer & Carlston, 1979) which in most conversations interact with content schema. Mary interprets Bill's comments about John by using her content schema for him, and by using her schema for Bill and his possible motives for telling her about John. In other words, conversants store motive schema along with content schema in order to facilitate the generation and interpretation of conversations.

When Bill mentions to Mary that John got a divorce, he is further directing Mary to focus on that aspect of her John schema that contains information about John's marital problem. As Mary retrieves and uses this aspect of her schema, she is also able to make inferences which explain for her the behavior of "she," or John's wife.

This schema theory analysis of the elliptical contents of our sample conversation generates one explanation of how the interactants created and assigned meaning to the contents of their conversation. Other analyses might focus more on the strategies the interactants were using to generate or interpret the contents of their utterances (for example, O'Keefe et al., 1980) or on both of these aspects, because they interact and occur together in conversations. By raising the John topic, Bill may be asserting that he knows something about John that Mary doesn't, as well as conveying new information about John. To focus exclusively on the possibility that Bill is stratgeically asserting his dominance over Mary captures only part of the sense Mary makes of Bill's comments.

Numerous schema theory studies of comprehension and recall have demonstrated the important role that topic or theme plays in making sense of the contents of written and oral messages (Bransford & Johnson, 1972; Bransford & McCarrell, 1974; Dodd & White, 1981; Dooling & Lachman, 1971; Housel, 1982; Lingle, Geva, Ostrom, Leippe, & Baumgardner, 1979). Results of these studies show that when a listener or reader is provided with an unambiguous topic or theme, he or she will have a greater likelihood of comprehending and recalling the contents of a message. From these studies, we know that providing the unambiguous topic or theme will help an interactant create an appropriate context for any conversational utterances because he or she will be signaling which schemata are appropriate to interpret a message's contents.

These studies of topic and theme also point out that when conversants or readers generate or are given an inappropriate topic or theme, they make inappropriate inferences, have trouble making sense of messages, and are likely to store and recall these inappropriate inferences. These findings are important in helping to identify problems interactants have in establishing similar social contexts for their conversations.

For example, Bill and Mary might discuss John's marital problem later. Mary could recall that Bill said that John got a divorce because he was running around on his wife. This topic, combined with Mary's elliptical inference that "she" (John's wife) was behaving the way she was because of John's infidelity, might create a slightly different context for this follow-up conversation from that of Bill's context. Bill recalls that he simply mentioned that John was getting a divorce. Whether Bill and Mary repair their differences, knowledge of their memory processes and of the changes in their schema for John lets the analyst explain the reasons for the changes in, or problems with, establishing a common social context in which to make sense of each other's comments.

As we get to know each other over time, we begin to anticipate these recall biases. Because Bill knows Mary well, and knows that Mary typically infers that when couples have marital problems the husband is running around on the wife, Bill can anticipate this recall bias in Mary's conversations. Bill may even choose to avoid certain topics because Mary will make certain unwarranted inferences, resulting in arguments between Bill and Mary after time passes or in tedious and time-consuming repair processes (see Snyder & Uranowitz, 1978, for further discussion of this recall bias phenomena).

The problem for the discourse analyst is that he or she is not always privy to the history of these recall biases, because the analyst usually studies fairly limited samples of interactants' conversation. One way to avoid this problem is to study the interaction of those who have no prior relationship. The analyst can then assume that the contents of their conversation will be free of these recall biases. While this type of interaction among strangers is fairly common, we cannot generalize from it to the equally common interaction that takes place between interactants with well-developed relationships. To study the way interactants who know each other well make sense of the contents of their conversation, the analyst must study their memories for shared information. This may require spending some time with the couples whose conversations are being studied to see what they normally do with the storage and retrieval of shared information. This extra time should also provide a better understanding of the subtle differences in the meaning of specific contents among different dyads, and of the fluid, ongoing, subtle process of conversational sense-making in unique social contexts. In the final analysis, this approach may lead to the identification of common mechanisms for creating and changing the meaning of content and, subsequently, to a better understanding of the processes that change social context.

Another phenomenon that affects the way conversants understand the contents of conversation is that conversants typically recall the inferences they make from conversational contents as well as, or in place of, the actual contents of a conversation (see

Clark & Clark's, 1977, reviews of this literature). Conversants often make inferences about the meaning of contents during the comprehension process (Fredricksen, 1975; Wanner, 1974). As time passes, these inferences about the meaning of conversational contents may bring about further inferences. This process creates reconstructive/constructive changes that are evident in an interactant's recall of a given conversation. The role of these inferential intrusions and of the inference-making process in general is the focus of the next section.

INFERENCE MAKING

We are all aware, intuitively as well as through observation of fellow interactants, that we constantly make inferences about the contents and motives/goals of those we talk with. If we did not make inferences, conversing would be an extremely time-consuming process, requiring that we make every intended inference explicit. Sometimes the inferences others expect us to make are too complex; we often ask them to make their inferences more explicit, although we may not ask for such explanations because we don't wish to reveal our ignorance. We assume that things will become clearer as we go, or we don't feel that the intended inference we missed was very important.

In our conversational sample, Mary has obviously made some inferences about John's wife and the motives for her behavior. Bill recognized and understood, or at least assumes that he recognized and understood Mary's inferences. If a researcher knew that Bill and Mary shared a memory of seeing John and his wife at a recent party and noticed that John and his wife were very sullen, that at one point John's wife screamed obscenities at him, and that she left the party immediately afterward, the researcher might assume that Mary's elliptical comments were inferences about John's wife being angry with him and that these comments became a part of Mary's schema for a couple getting a divorce. Drawing upon schema theory, the analyst might assume that Bill and Mary had a schema for John and his wife's relationship, or more specifically, their marital problem, and that this schema contained Bill and Mary's mutual knowledge of John and his wife's strange/unfriendly behavior at the party. Given that Bill and Mary shared this information, the analyst might assume that they had similar general schemata for getting a divorce in this culture, and that by drawing on this more general schema (for divorce) and retrieving specific information (the couple's fight at the party) from the "John-and-his-marital-problem" schema, Mary could retrospectively infer a motive for John's wife's behavior. That Bill acknowledged the inferences as correct would lead the analyst to assume that Bill had made the same inferences as Mary from the wider possible array of inferences that outside observers might have made. The analyst could check his assumptions by conferring with the interactants, or possibly by observing the contents of further interactions between Bill and Mary on this topic.

The problem with relying on observing further interactions is that Bill and Mary might not explicitly raise the topic of John's marital problems again, making a direct check of the accuracy of Bill's interpretation of Mary's inferences impossible. Because Bill and Mary never discussed John's wife's behavior directly, we cannot be sure that Mary is making inferences about her motives for behaving the way she did, or that Bill is making the same inferences. Therefore, it makes sense for the analyst, in this case, to check his or her assumptions about the inferences Bill and Mary are making by asking Bill and Mary.

If we find Bill and Mary discussing the marital problems of a different friend who is bringing his wife to a party that Bill and Mary will also attend, Bill might say to Mary, "I hope they don't create a scene," and Mary might reply, "We both know what can happen when people get divorced." A conversational analyst who focused on this conversation's contents, knowing nothing about this couple's shared experiences, might infer that they are discussing the fact that divorcing couples can be hostile toward one another in public gatherings. While this inference is certainly warranted and even correct at a general level of analysis, it misses the potentially richer interpretation of the meaning of this conversation's contents—namely, assumptions about what Bill meant by "creating a scene" and what Mary meant by "what can happen."

Harris's (1981) review of the inferential process indicates that there are different kinds of inferences interactants/receivers can make, some of which are more warranted than others. Mary's inference that John's wife behaved in a hostile manner because John was divorcing her was less warranted than Mary's inference that Bill was talking about the John they both knew who was having marital problems. But we can make this inference only if we know enough about Bill and Mary's shared memories to know that they only have one friend named John who is having marital problems.

Harris's review also acknowledges the important role of themes and the schemata that themes help interactants retrieve from memory. The inferences conversants make from the contents will depend, to a large extent, on the themes and subsequent schemata they use to make sense of a given conversation. For example, Housel (1982) found that giving subjects unambiguous themes for conversations caused them to make more accurate inferences in delayed free recalls of conversations than when they were provided with ambiguous themes for the same conversations. In our sample conversation, it makes sense that Bill, by providing a divorce theme for a specific couple (John and his wife), gave Mary the memory schema retrieval instructions she needed to make the inferences she did about why John's wife behaved as she did at the party.

Many researchers (for example, Bransford & McCarrell, 1974; Owens et al., 1979) have referred to the schemata subjects were assumed to have used in interpreting stimulus messages as the context for the messages. Using the notion of topic/theme and the presumed link to schema retrieval, we can begin to understand why some conversational inferences are more accurate and warranted than others. These conversations, for which interactants can develop unambiguous themes and retrieve specific schemata representations of shared memories, allow or motivate the conversants to make more accurate and warranted inferences. These inferences are limited by the degree to which the interactants have developed specific schemata for given shared memories. The more general or abstract the schemata, the more likely that a greater variety of inferences will be possible.

In analyzing conversational contents for inference, the analyst should attempt to isolate and define the schemata that interactants are relying on to make inferences, as well as the level of specificity or detail of the schemata. The conversations alone may provide all the detail an analyst needs to accomplish these goals, but as the sample conversation and its proposed continuation should make clear, not all inferences are made explicit in conversation; many must be assumed. This is especially true when the conversants piggyback inferences one on top of another. For example, Mary inferred that John's wife caused a scene at the party because he was divorcing her, and from this inference Mary erroneously inferred that John was running around on his wife. Mary never made these inferences explicit in conversation, but the inferences subtly changed her schema for the divorce process, as

well as her schema for John. The result of these unstated or implicit inferences might show up in some permuted form in a later
conversation with Bill.

In sum, interactants make implicit inferences, recall these inferences, change the state of their world knowledge to accommodate these inferences, and use these inferences to create new
content meanings. Thus they partially create and change the social
context within which they communicate. Studying conversational
contents without reference to this subtle and pervasive inference-
making process may limit the analyst's understanding of the way
individuals create and interpret new meanings for conversational
contents. Only by making the effort to get to know interactants
over time can the conversational analyst determine typical
inference-making and recall patterns that interactants rely on to
create and interpret new meanings for the contents they use to
converse.

STRUCTURE

Structures play a role in helping interactants make sense of conversations. Speakers use a variety of structures—some mental,
others realized in conversation—to help them generate and
understand conversations.

In our sample conversation, Bill and Mary take turns at talking. Bill opened the conversation, and Mary closed it. Were Bill
and Mary following some set of conversational rules (such as turn-
taking) while conversing? O'Keefe (1979) described two positions
ethnomethodologists take on this issue: (1) It does not matter what
Bill and Mary are thinking about (for example, whether to follow
certain patterns of conversational behavior). What matters is what
they do, or what they can be observed doing in the conversation
(Schegloff & Sacks, 1973). (2) The cognitive processes that Bill
and Mary are using during the conversation are the keys to
understanding the structure of their interaction (Cicourel, 1980).

Rather than debate the merits of these two approaches to structure, I want to demonstrate some ways a conversational analyst
might approach this structure issue. Because I have argued that
we need to look at the contents which are to be found in interactants' memories, I will focus on those aspects of structure that have
cognitive correlates. Schema theory offers some promising approaches to this issue of how structure helps conversants make
sense of conversations.

As linguistic research has demonstrated, and as those who teach writing know, single sentences can take a limited range of syntactic structures. In conversations, however, utterances can take a broader range of syntactic structures (Clark & Clark, 1977). Using the story or passage as the unit of analysis, recent schema theory research (for example, see Rumelhart, 1975) has proposed a series of story structures or syntaxes. Dillon (1981, p. 51) summarized this work when he proposed structural schema that "give form or structure to experience." Schema theorists argue that interactants or readers use these structural schemata as templates to organize and comprehend stories, passages, jokes, or conversations. Other research (for example, Adams & Bruce, 1980; Steinberg & Bruce, 1980) has demonstrated that these expectations are so strong for story structures that when readers are given stories in which structural units (introduction, conflict, conflict resolution, and so on) are in random order, readers will remember the stories in their structural schema-predicted order. Similarly, Ellis, Hamilton, and Aho (1983) found that when subjects are given the transcript of a conversation in which statement pairs have been randomized, subjects rearrange these statement pairs in a predictable order based on their assumed knowledge of how conversations should be structured.

While these structural expectations are very strong, interactants or readers can make sense of poorly (that is, not in accord with expected structural schemata) organized stories or conversations if they are very familiar with the contents of the message.

"Good content schema will often carry the day: one can paragraph "rhetorically," omit transitions, leave topic sentences implicit, and on through the whole list of ways that real texts deviate from the prescription of what one should do.... The mistake often made, however, is to think that the formal structure is weak at the points where these needs arise: in fact, it may be the content schema that has run out (or broken down) or is understood in a different way [Dillon, 1981, p. 55].

This is true for those conversations we have with others that follow normal interactional structure, but whose contents are so general or ambiguous that we have great difficulty understanding the interactant. Dillon (1981) further argues that structural schema that match interactants' or readers' expectations for the way a message will be structured can aid them in making sense of messages, as long as the messages have adequate content.

The significant point Dillon makes, indirectly, is that structure and content interact in the sense-making process. The reader has probably been involved in conversations where a topic is raised on which an interactant is known to be an expert. As a result, other interactants may give the "expert" interactant longer turns and expect him or her to interrupt others on this topic. When we talk to another about his or her dearly departed relative, we may expect and allow longer pauses in our conversation, and if the other does speak, we are likely to give them great leeway in turn length. These are just two examples of how the contents or the topic interact with the structure of a conversation. If the analyst chooses either content or structure exclusively as the focus of the study, he or she is bound to miss this content-structure interaction and is therefore less likely to understand how conversants make sense of conversations in part by relying on both of these variables.

In studying conversational structure, the analyst must realize that there are levels of structures, just as there are levels of social context, contents, and inferences. Supporting this assertion, Dillon (1981, p. 51) explains: "Schema come in varying levels of generality and operate in different ways." Likewise, Clark and Clark (1977, p. 236) observe: "Although discourse has structure—conversations and description attest to that—it has two kinds of structure: hierarchical and logical. . . . No matter whether the discourse is a dialogue or monologue, speakers have an overall hierarchical plan that they try to abide by throughout. Yet each sentence (or utterance) has to be planned locally." O'Keefe et al. (1980) also use these notions of hierarchical and local planning to analyze conversations. By acknowledging that interactants approach conversations as problem-solving tasks where they must try to plan their utterances to fit into some overall plan for the conversation, the analyst may begin to understand why interactants select given contents, why they interrupt or permit interruptions, why they change topics, and so on. Of course, not all speakers are sufficiently strategic or have the conversational skills to effectively implement a hierarchical plan, but by knowing or inferring their plans, the analyst can begin to understand why they structure their utterances (and on a more global level, their conversations) the way they do.

To better understand the role of structure in making sense of conversations, analysts should recognize that content interacts with structure, and that varying levels of structure exist in conversations. The analyst should also heed Clark and Clark's (1977) observation: "Discourse also varies in how much of its structure is conventional and how much is planned anew each time by problem solving." Studying couples interacting over time, the analyst can begin to observe characteristic structures in their conversations on given topics. For example, there are typical structures for dialogues on given topics (having children, buying a new car, and so on). Their patterns of interaction (such as turn-taking length) and the contents of their utterances are surprisingly predictable, and there is even some predictability across couples on given topics. The analyst can observe these regularities by watching and interacting with specific couples over time and sharing memories with them. Clearly, the conversational analyst must observe and question interactants over time if he or she wishes to understand how individual conversants make sense of conversations.

CONCLUSIONS AND RECOMMENDATIONS

Conversational analysts should study conversants' mutual memories in order to understand how they create and make sense of contents. Specific dyads should be studied longitudinally as a way to monitor conversants' shared memories as they help to create and define social contexts over time.

Memory's role. As conversants, we intuitively recognize that we share memories with imtimates that are more detailed and dyad-specific than memories we share with others in our culture. We also recognize intuitively that we could not converse on given topics without recalling information pertinent to those topics. As conversational analysts, we should investigate these intuitions, since they are so pervasive in the conversational process. While this would add more complexity to the typical conversational analysis, we cannot afford to ignore this crucial aspect of the conversational process.[3]

There are a variety of memory metaphors available in the current research which can help conversational analysts reduce the complexity of this issue (for example, scripts, frames, and schemata). Analysts could develop a memory metaphor based on texts of conversation and excluding any reference to mental representations within conversants. This would solve the potential problems of unreliable or inaccurate subject self-reports, but this approach would ignore the large body of memory research demonstrating that interactants recall inferences from conversations but only rarely exact contents. At some point the analyst would have to develop a memory representation that would account for these inferences.

Schema theory is a promising metaphor, because it allows the analyst to focus on the structures and contents of conversants' shared memory and to account for the inferences that are stored in memory. Conversants' mutual memory schemata could be mapped in terms of their contents, the relations between their contents, and the likely inferences based on the contents' structural relationships, as well as in terms of errors in recall, biases in recall, and the ways conversants change their mutual schemata by integrating new information. A schema theory analysis would allow the analyst to follow and describe these processes over time. This approach might result in a better understanding of how specific dyads create and interpret conversational contents. Future schema research into the conversational memory process must focus on how conversants generate utterances and conversations using their mutual schemata, and other nonmutual schemata, in ongoing conversations (Chafe, 1977).

Memory processes must be taken into account, because conversants carry information from one conversation to the next for functional reasons—to avoid having to generate completely new contexts each time they converse. In other words, carrying information forward to new conversations helps interactants establish unambiguous context quickly and easily.

Studying inferences. The conversational analyst needs some way to limit the potential range of plausible inferences a speaker makes. The analyst's cultural conversational competence will allow him or her to make reasonable guesses as to what inferences are possible. However, the analyst who wishes to study inferences at a lower level of abstraction (such as intimate dyads) could gather conversants' self-reports and compare them with the historical records of the conversants' interactions.

Schank (1983) argues that "scripts" (a permutation of schema theory) are a useful way to approach the study of inference making. He found that in developing a computer program that simulated human interactions,

> the idea of scripts allowed the computer to make more intelligent inferences because it (computer) knew about the domain it was operating in. In its car-accident script, for example, we built in the police investigation and the hospital and the ambulance so that it

would assume that the ambulance had come if we told it that John had been in a car accident and had been taken to the hospital [Schank, 1983, p. 32]

As conversants, we recognize that we make numerous inferences during most conversations. As conversational analysts, we must develop methods to describe and understand this intuitively obvious process; the schema theory approach has proven fruitful in this regard. Schema theory offers a way to achieve specificity content-wise and to give structure to the concept "world knowledge," which until now has been represented as an ambiguous, amorphous repository of all our stored knowledge.

Focus on specific dyads. Focusing on specific dyads and following their conversational history allows the conversational analyst to observe the subtle structural and content changes that characterize the shifts in context and meaning which typify intimate relationships. Focusing on the intimate dyad, an analyst can begin to delimit the scope of the interactants' social world and the contents of their world knowledge, thus enabling the investigator to analyze a limited set of data and to increase the power of his or her explanations. As conversational analysts, we can avoid the extreme of knowing very much about very little (for example, single-word or utterance explanations). Studying the intimate dyad, the analyst can focus on the whole conversational level in order to make detailed observations about content and structural changes.

Obviously, an extremely lengthy essay would be required to review adequately all the various approaches currently used in the study of social context, conversational contents, structure, and inference processes. The purpose of this chapter was not to attempt such a comprehensive review, but to sketch some potentially fruitful ways of conceptualizing and studying these issues. It is hoped that this purpose has been served through an albeit brief review of schema theory and a somewhat abbreviated list of suggestions for future research.

NOTES

1. A partial list of recent communication articles includes Craig (1979), Goss (1982), Housel (1982), Beach (1983), Hawes (1977), Housel and Acker (1979), O'Keefe and Delia (1982), Planalp (1983), Stafford and Daly (1983), and Smith

(1982). A partial list of ethnomethodologists working in this area is Sacks, Schegloff, and Jefferson (1974), Schegloff and Sacks (1973), Cicourel (1980), and Rice (1980). A brief list of psychologists working on this issue is Keenan, MacWhinney, and Mayhew (1977), Spiro (1980), Clark and Clark (1977), Kintsch and van Dijk (1978), Fredericksen (1975), and van Dijk (1977). A partial list of linguists is Chafe (1977), Fillmore (1977), Tannen (1979), Bennett (1981), Gumperz (1982), Lakoff (1982), De Beaugrande (1980), Freedle (1979), Dillon (1981), and Reichman (1978). Several artificial intelligence researchers have also been investigating the conversational meaning-making process: Winograd (1977), Schank and Abelson (1977), Schank (1983).

2. In light of O'Keefe's (1979) observation that "the ways in which the public, observable features of behaviour produce a sense of social order is manifestly not 'subjective' " (p. 209), the definition of "public" and the implication of nonsubjectivity bears further consideration and investigation.

3. Recently, Schank (1983) argued that "there is no discussing language without discussing memory. They are completely tied together" (p. 32). Schank also advocated the use of scripts, a permutation in schema theory research, as a way to constrain the possible inferences interactants would make about the meaning of conversational contents. "It [the inference-meaning process] had to be constrained in some way. Scripts were a solution to the problem."

REFERENCES

Adams, M., & Bruce, B. (1980). *Background knowledge and reading comprehension* (Reading Education Report No. 13). Urbana: University of Illinois, Center for the Study of Reading.

Bartlett, F. (1932). *Remembering: A study in experimental and social psychology.* Cambridge: Cambridge University Press.

Beach, W. (1983). Background understandings and the situated accomplishment of conversational telling-expansions. In R. Craig & K. Tracey (Eds.), *Conversational coherence: Studies in form and strategy.*

Bennett, A. (1981). Interruptions and the interpretation of conversation. Discourse Processes, 4, 171-188.

Bransford, J., & Johnson, M. (1972) Contextual prerequisites for understanding: Some investigations of comprehension and recall. *Journal of Verbal Learning and Verbal Behavior, 11,* 717-726.

Cantor, N., & Mischel, W. (1979). Prototypes in person perception. In L. Berkowitz (Ed.), *Advances in experimental social psychology.* New York: Academic Press.

Chafe, W. (1977). The recall and verbalization of past experience. In R. Cole (Ed.), *Current issues in linguistic theory.* Bloomington: Indiana University Press.

Chomsky, N. (1965). *Aspects of the theory of syntax.* The Hague: Mouton.

Cicourel, A. (1980). Three models of discourse analysis: The role of social structure. *Discourse Processes, 3,* 101-132.

Clark, H., & Clark, E. (1977). *Psychology and language: An introduction to psycholinguistics.* New York: Harcourt Brace Jovanovich.

Cohen, C., & Ebbesen, E. (1979). Observational goals and schema activation: A theoretical framework for behavior perception. *Journal of Experimental Social Psychology, 15,* 305-329.

Craig, R. (1979). Information systems theory and research: An overview of individual information processing. In D. Nimmo (Ed.), *Communication Yearbook 3.* New Brunswick, NJ: Transaction Books.

De Beaugrande, R. (1980). *Text, discourse, and process: Toward a multi-disciplinary science of texts.* Norwood, NJ: Ablex.

De Villiers, P. (1974). Imagery and theme in recall of connected discourse. *Journal of Experimental Psychology, 103,* 263-268.

Dillon, G. (1981). *Constructing texts: Elements of a theory of composition and style.* Bloomington: University of Indiana Press.

Dodd, D., & White, R. (1981). *Cognition: Mental structures and processes.* Boston: Allyn Bacon.

Dooling, D., & Lachman, R. (1971). Effects of comprehension and retention of prose. *Journal of Experimental Psychology, 88,* 216-222.

Ellis, D., Hamilton, M., & Aho, L. (1983). Some issues in conversation coherence. *Human Communication Research, 9,* 267-282.

Fillmore, C., (1977). Topics in lexical semantics. In R. Cole (Ed.), *Current issues in linguistic theory.* Bloomington: Indiana University Press.

Fredericksen, C. (1975). Effects of context-induced processing operations on semantic information acquired from discourse. *Cognitive Psychology, 7,* 139-166.

Freedle, R. (1979). *New directions in discourse processing.* Norwood, NJ: Ablex.

Goss, B. (1982). *Processing communication: Information processing in intrapersonal communication.* Belmont, CA: Wadsworth.

Gumperz, J. (1982). *Discourse strategies.* Cambridge: Cambridge University Press.

Harris, R. J. (1981). Inferences in information processing. In G. Bower (Ed.), *The psychology of learning and motivaiton.* New York: Academic Press.

Hastie, R., Ostrom, T., Ebbesen, E., Wyer, R., Hamilton, D., & Carlston, D. (1980). *Person memory: The cognitive basis of social perception.* Hillsdale, NJ: Lawrence Erlbaum.

Hawes, L. (1977). *Suggesting and accepting: A conversational practice and some of its structural variations.* Paper presented at the Speech Communication Association Convention, Washington, DC.

Housel, T. (1982). *Conversational processing: The effects of theme and attention focusing strategy on comprehension, recall accuracy, and uncertainty reduction.* Paper presented at the Western Speech Communication Association, Denver.

Housel, T., & Acker, S. (1979). *"Schema theory": Can it connect communication's discourses?* Paper presented at the International Communication Association Convention, in Philadelphia.

Keenan, J., MacWhinney, B., & Mayhew, D. (1977) Pragmatics in memory: A study of natural conversation. *Journal of Verbal Learning and Verbal Behavior, 16,* 549-560.

Kintsch, W., & van Dijk, T. (1978). Toward a model of text comprehension and production. *Psychological Review, 85,* 363-394.

Lachman, R., Lachman, J., & Butterfield, E. (1979). *Cognitive psychology and information processing: An introduction.* Hillsdale, NJ: Lawrence Erlbaum.

Lakoff, R. T. (1982). Persuasive discourse and ordinary conversation, with examples from advertising. In D. Tannen (Ed.), *Analyzing discourse: Text and talk* (Georgetown University Roundtable on Languages and Linguistics). Washington, DC: Georgetown University Press.

Lingle, J., Geva, N., Ostrom, T., Leippe, M., & Baumgardner, M. (1979). Thematic effects of person judgments on impression organization. *Journal of Personality and Social Psychology, 37,* 674-687.

O'Keefe, B., & Delia, J. (1982). Impression formation and message production. In M. Roloff & C. Berger (Eds.), *Social cognition and communication.* Beverly Hills, CA: Sage.

O'Keefe, B., Delia, J., & O'Keefe, D. (1980). Interaction analysis and the analysis of interaction organization. In N. Denzin (Ed.), *Studies in symbolic interaction: Vol. III.* Greenwich, CT: JAI Press.

O'Keefe, D. (1979). Ethnomethodology. *Journal of the Theory of Social Behavior, 9,* 187-217

Owens, J., Bower, G., & Black, J. (1979). The "soap opera" effect in story recall. *Memory and Cognition, 7,* 185-191.

Planalp, S. (1983). *A test of the impact of three levels of relational knowledge on memory for relational implications of messages.* Paper presented at the International Communication Association, Dallas.

Reichman, K. (1978). *Conversational coherency* (Technical Report No. 95). Urbana: University of Illinois, Center for the Study of Reading.

Rice, G. (1980). On cultural schemata. *American Ethnologist, 7,* 151-171.

Rumelhart, D. (1975). Notes on schema for stories. In D. Bobrow & A. Collins (Eds.), *Representation and understanding: Studies in cognitive science.* New York: Academic Press.

Sacks, H., Schegloff, E., & Jefferson, G. (1974). A simple systematic for the organization of turn-taking for conversation. *Language, 50,* 696-735.

Schank, R. (1983). A conversation with Roger Schank. *Psychology Today,* April, 28-36.

Schank, R., & Abelson, R. (1977). *Scripts, plans, goals, and understanding.* Hillsdale, NJ: Lawrence Erlbaum.

Schegloff, E. (1968). Sequencing in conversational openings. *American Anthropologist, 70,* 1075-1095.

Schegloff, E., & Sacks, H. (1973). Opening up closings. *Semiotica, 8,* 289-327.

Smith, M. J. (1982) *Persuasion and human interaction: A review and critique of social influence theories.* Belmont, CA: Wadsworth.

Snyder, M., & Uranowitz, S. (1978). Reconstructing the past: Some cognitive consequences of person perception. *Journal of Personality and Social Psychology, 36,* 941-950.

Spiro, R. (1980). Accommodative reconstruction in prose recall. *Journal of Verbal Learning and Verbal Behavior, 19,* 84-95.

Stafford, L., & Daly, J. (1983). *Conversational memory: The effects of recall mode and instructional set on memory for naturally occurring conversations.* Paper presented at the International Communication Association Convention, Dallas.

Steinberg, C., & Bruce, B. (1980). *Higher-level features in children's stories: Rhetorical structure and conflict* (Reading Education Report No. 18). Urbana: University of Illinois, Center for the Study of Reading.

Tannen, D. (1979). What's in a frame? Surface evidence for underlying expectations. In R. Freedle (Ed.), *New directions in discourse processing.* Norwood, NJ: Ablex.

van Dijk, T. (1977). Semantic macro-structures and knowledge frames in discourse comprehension. In M. Just & P. Carpenter (Eds.), *Cognitive processes in comprehension.* New York: John Wiley.

Wanner, E. (1974). *On remembering, forgetting, and understanding sentences.* The Hague: Mouton.

Winograd, T. (1977). A framework for understanding discourse. In M. Just & P. Carpenter (Eds.), *Cognitive processes in comprehension.* New York: John Wiley.

Wyer, R., & Carlston, D. (1979). *Social cognition, inference, and attribution.* Hillsdale, NJ: Lawrence Erlbaum.

9

Some Functions of Feedback in Conversation

Robert Kraut and Steven H. Lewis

Coordination is the central problem in conversation (Grice, 1975; Kraut & Higgins, in press). In planning all aspects of discourse, fron structure to word choice, speakers need to shape what they have to say to their audience. In order to achieve their social goals with a particular audience, speakers must plan their speech with that audience in mind. They make inferences about particular mental states of their audiences. These inferences are based on their prior knowledge of the audience, on many of the audience's observable social characteristics, and on the situation in which the speakers and listeners find themselves.

Adjusting one's speech to a target is not simply a matter of decorating a preformed and asocial message. Instead, as Mead

Authors' Note: *This research was supported by grants from the National Institute of Mental Health and the National Science Foundation to the first author. An earlier version of this chapter was presented at the meeting of the American Psychological Association, Montreal, September 1980. Requests for reprints should be sent to Robert Kraut, Bell Communications Research, Inc., 600 Mountain Avenue, Murray Hill, NJ 07974.*

(1934) argued, this ability lies at the core of communication and at the core of being social. Communication requires speakers' ability to infer what listeners already know so that the speakers can correctly phrase the messages they provide. As Clark (in press) has argued, adjustment or coordination—"the idea that speakers attempt to coordinate what they say with the interpretations their addressees impute to what they say"—"is central at all levels of language, from individual words to conversations and narratives."

The concept of coordination is so central to communication that researchers may be tempted to dismiss it or overlook it because of its obviousness. "Of course," one might protest, "people cannot communicate unless they agree on a common language, with a common vocabulary and syntax." We argue, however, that the challenge is to go beyond Mead's proclamation of the centrality of coordination and to trace both the mechanisms through which people manage to coordinate communication and the cognitive and social consequences of such coordination.

People coordinate their communication through both premeditation and nonpremeditatated interaction. In a premeditated mode, when planning their speech speakers take into account their knowledge of the language, the topic they are addressing, and the characteristics of their listeners, including what they already know. In an interactional mode, speakers also adapt what they are saying to their own prior speech and to their partners' contributions. Each of a speaker's conversational subgoals can be planned noninteractively, as speechwriters and novelists do, relying solely on their general knowledge of the topic at hand and the intended audience, without direct feedback from that audience. On the other hand, plans can be modified during the course of a conversation on the basis of explicit feedback from conversational partners.

The traditional literature in communication has often failed to pay attention to the ways in which people coordinate their contributions (for example, see McGuire, 1969, on attitude change, and Rosenthal, Hall, DiMatteo, Rogers, & Archer, 1979, on nonverbal communication). Even work that has emphasized coordination (for example, Hale & Delia, 1976) has ignored interactive coordination (but see Labov & Fanshel, 1977).

What these researchers ignore is the almost continuous feedback that speakers receive from their listeners' verbal, paralinguistic, and nonverbal behavior. This feedback provides them with information about the listeners' cognitive and affective states which helps speakers form and reform their speech. In addition to planning and forming their speech, speakers are engaged in on-line judgments of their partners' transient states that have immediate consequences for their own behavior. Although speakers can plan conversations, both planning and execution may lead to communication problems. The difficulties of simultaneously planning future speech, executing current speech, and evaluating past speech mean that speakers frequently cannot catch all their miscalculations and mistakes through self-evaluation. By using feedback from actual listeners, speakers can modify their plans and repair their execution based on evidence of the consequences of their speech.

We can distinguish between two types of feedback that audiences provide speakers. First is the semantically meaningful speech that listeners provide when they take up the speaking turn. Almost everything that one speaker says in response to another informs the original speaker about the listener's understanding of, agreement with, and interest in the original remark. These responses may not be intended to provide feedback; the simple requirement that conversation be relevant to the context and the topic at hand, however, means that a speaker can often glean a listener's reaction from the listener's response.

Speakers also rely on "back channel" responses (Yngve, 1970) to gauge their audience's reaction. Listener responses or back channel communications are the small visual and auditory comments an auditor makes while a speaker is talking, without taking over the speaking turn. Unlike general nonverbal reactivity (such as eye-contact) or full semantic responses (for example, listeners taking over the speaking turn), back channel responses seem to be specialized communication devices that provide speakers with feedback during conversation. They occur frequently in conversation and seem to have little function other than to inform a speaker of a listener's mental state. Back channel responses include clarifying questions, brief verbal responses (such as "yeah," "mm-hmm," and the like), head nods, brief smiles, repetition of the speaker's words, and brief sentence completions (Duncan, 1974). The boundary, however, between them and actual exchanges of a speaking turn is not sharp.

Although many writers have speculated that variations of listener responses indicate to speakers the degree to which listeners are paying attention, understanding points being made, or agreeing with them (for example, Yngve, 1970; Rosenfeld,

1978), research to date has been more suggestive than definitive. For example, Leathers (1979) suggested that judges of verbal and nonverbal feedback signals can reliably interpret them as messages concerning the listener's involvement, state of confusion, thoughtfulness, and feelings about what had previously been said. Kendon (1967) has shown that speakers look away from listeners as they start a cognitively difficult passage, but return their gaze to the listeners toward the end of the passage, as if calling for their reactions. Listeners tend to give back channel responses at these points (Dittmann & Llewellyn, 1968; Duncan, 1974).

Some research has examined whether feedback in conversation influences the quality of speech, as would be expected if one of its functions is to coordinate understanding between speaker and listener. When speakers are denied feedback, they become upset and disrupted, and their speech becomes less structured and coherent (Kent, Davis, & Shapiro, 1978). At least by some measures, their speech becomes less efficient (Krauss & Bricker, 1967; Krauss, Garlock, Bricker, & McMahon, 1977; Krauss & Weinheimer, 1964, 1966). Krauss and his colleagues used a referential task in which speakers described difficult-to-name objects to listeners. When back channel communication to speakers was disrupted, speakers used more words to describe the objects successfully. While most speakers started with lengthy descriptions that they abbreviated over time, speakers lacking feedback did not abbreviate as much, presumably because they received no confirmation that their partners had understood their shorthand.

The purpose of the research presented here is to examine how listener responses influence the outcome of conversation. This was done by examining some of the ways such responses influence the content of speech. In this research, we first asked whether the amount of listener feedback would influence a speaker's performance on two conversational tasks that were more complex and more natural than the highly structured referential tasks used by Krauss and his colleagues. In parallel analyses, we examined the changes in speech produced by listener responses that might have mediated these changes in performance.

In one task speakers lied to naive interviewers, and in the other they summarized a movie to listeners. Both are natural tasks with an objective outcome that structures the topics discussed while still providing subjects with considerable conversational latitude. They differ in that deception is primarily a persuasive task in which the deceiver and the audience often have opposing motives, while summarization is primarily an information exchange task in which the speaker and audience share motives. Both, however, involve important social skills.

As in previous research, we studied the role of listener responses by comparing the performance of speakers who received normal listener feedback with those who received less than normal feedback. We reasoned that to the extent that speakers monitor their partners' feedback in order to assess whether they have understood and agreed with particular points, speakers will modify their speech to take into account their partners' level of understanding and objection. What the speakers say and what the partners need to know should be more coordinated in conversations in which the partners are providing adequate feedback. Given the conflicting motives of deceivers and detectors of deception, in the deception experiment this coordination may help either the deceivers, making them better liars, or the detectors of deception, making the liars less successful. The case of summarization is more straightforward. Summaries with increased coordination should be better, in the sense that listeners should understand more about the movie.

The results, described in more detail below, show that even a small amount of back channel feedback influences conversational outcomes. In both the lie detection and the summarization experiments, feedback aided listeners, including both the person who originally provided the feedback and others who read about or eavesdropped on the conversation without providing feedback. The actual providers of feedback were aided more than eavesdroppers, however.

To show that feedback coordinates an interaction between speaker and listener, we also need to examine in more detail the relationships among listeners' feedback, listeners' mental states (for example, whether they are confused about what the speaker is saying), and speakers' subsequent behavior (for example, whether they repeat material or move on to new topics).

We looked at these relationships in different ways in the deception and summarization experiments. In the summarization experiment, we manipulated listeners' understanding and then examined the effect on their use of back channel responses and on speakers' presentation of information to listeners. In the deception experiment we examined the relationship between listener feedback and the semantic structure of discourse. Let us assume for the moment that listener feedback communicates an "I understand" signal to speakers. If so, we would expect that listeners would provide more feedback as they understood the conversation better. If speakers are denied information about when their listeners understand, they may be tempted to provide listeners with too much detail. When they get an "I understand" signal, they are more likely to close up the subtheme they were developing and continue on with the main theme or story line. These are the sorts of hypotheses we tested in the two experiments.

FEEDBACK AND DECEPTION

To see how feedback influences the ability to lie, we conducted two sets of interviews one year apart, one with college-student subjects (twelve groups) and the other with noncollege-student, adult subjects (thirty groups) as speakers and interviewers. In both, a speaker-subject was interviewed sequentially by two interviewer-subjects. The speaker was instructed to lie on half the answers. One interviewer was instructed to be passive, providing no back channel or affective feedback to the liar. The active interviewer was given no instructions and provided the liar with normal back channel responses.

To assess the impact of feedback on the quality or outcome of speech, we had new judge-subjects try to guess on the basis of transcripts when the original speakers had been lying. To assess the impact of feedback on the structure of speech, we also coded these transcripts for their semantic structure.

Feedback and Liars' Effectiveness

Methods

The interviews. The speakers' task was to answer questions about their personal history and opinions that the interviewers would ask them. They were instructed to make all their answers as convincing as possible while lying on a predesignated half of them. They were given the additional incentive of a twenty-dollar prize if they were the most effective liar.

The interviewer subjects' task was to get to know the interviewee by asking him or her some questions provided. In addition, the passive interviewers were told: "Your task is to be a neutral and passive interviewer, to provide the interviewee no feedback on his or her answers. . ." They were given specific instructions about behavior to control (for example, headnods, smiles, and "mm-hmms") in order to be passive. In contrast, the active interviewers were encouraged to "act as naturally as possible." They were given no other instructions about their verbal and nonverbal behavior.

Judgment task. In order to discover whether the responsiveness of a conversational partner influences how well speakers lie, we had a sample of judge-subjects judge the truthfulness of pairs of active and passive answers from the original interviews. Differences in their accuracy in judging deception based on these paired answers were expected to reflect directly the quality of the speakers' performances in talking to active and passive interviewers. (If we had the original interviewers make the judgments, their accuracy could have been a direct function of both their behavior and the effects their behavior had on the liars' answers.) Judges made their decisions after reading a transcript of question-and-answer-sequences. The transcripts were edited to eliminate differences between active and passive interviewers (that is, listener responses from active interviewers were removed).

We had two samples of judges make truthfulness judgments. One sample of 45 college students judged only the transcripts from the first replication. The second sample of 73 college students judged the transcripts from both replications. Judges were given a description of the way in which the transcripts had been made. They were not told about the interviewer activity manipulation, but were told that the base rate of lying was about 50 percent. They rated the truthfulness for each answer they read on a six-point scale, where 1 indicated that they were confident that the speaker was lying, and 6 indicated that they were confident that he or she was telling the truth.

We standardized the truth judgments so that each judge had a mean of zero and a standard deviation of one (see Cronbach, 1955), controlling for differential elevation and spread. We then computed an accuracy score from these standardized truth judgments by adding the absolute values when judges identified a lie or truth accurately and subtracting them when they were inaccurate. Finally, we computed the mean accuracy for each excerpt, where each mean was based on judgments from eight to ten judges. The mean of each question was the basis for the analyses to follow, and the N for each analysis was the number of excerpts; that is, 24 pairs from 12 speakers in the first replication and 30 pairs from 30 speakers in the second replication.

Liars' Effectiveness

Our hypothesis was that speech to an active interviewer should be better coordinated with what a listener needs to know, leading to better lying if the resulting clarity differentially aids the liar, or to worse lying if the clarity differentially aids a detector of deception. We asked whether judges were better able to detect deception from active or passive interviews, once for the first sample and once for the second sample. For the second sample, we also asked whether the replication from which the transcripts came (that is, adult versus college student samples) and the replication by listener responsiveness interaction had significant effects.

In both samples, judges reading transcripts from active interviews were more accurate (that is, the liars were poorer) than when they read transcripts from passive interviews involving the same speakers (see Table 9.1). Combining the significance levels for independent samples of judges, $z = 1.86$, $p = .06$. In the second sample, tests of the main effect of the replication from which the excerpts came ($p > .20$) and the replication's interaction with the interviewers' responsiveness ($p > .80$) did not approach significance.

Why was the speech from speakers talking to active rather than passive partners easier to see through? If the presence of a responsive listener increased the coordination between a speaker and a conversational partner and thus produced a clearer and more comprehensible answer, as we have assumed, the greater comprehensibility of these answers may have highlighted content cues to deception, such as implausibility, inconsistency, vagueness, or extraneous detail (see Bolinger, 1973; Hemsley, 1977; Kraut, 1978, 1980; Kraut & Poe, 1981; Zuckerman, DePaulo, & Rosenthal, 1981). These cues to deception, which can occur in perfectly lucid prose, would have made the speakers talking to active interviewers worse liars when judgments were being made from transcripts.

Listener Feedback and Semantic Structure

In designing this research, we hypothesized that speakers would change their speech in response to feedback from active partners about their attention, comprehension, and agreement. An examination of speakers' answers when talking to active and passive interviewers should give some insight into the way feedback causes people to modify their speech.

TABLE 9.1   Accuracy of Raters' Judgments of Deception According
to Feedback Provided by the Original Interviewer

| | Interviewer Activity | |
Replication	Active	Deadpan
Replication one	.18	.04
Replication two	.21	.13

NOTE: Accuracy scores are based on truth judgments standardized by rater. Higher numbers indicate greater accuracy in these standard deviation units. Zero indicates chance accuracy.

To see how speakers with active and passive partners organized their speech, we adapted Grimes's (1975) analysis of rhetorical predicates to our interviews. Grimes's system is an attempt to specify the organization of discourse by specifying the semantic relationships betweeen the ideas in it. Since back channel responses often occur between grammatical or phonemic clauses (Duncan, 1974; Dittmann & Llewellyn, 1968), this coding system based on the semantic relations between clauses is especially appropriate.

Procedure

We will describe the mechanics of applying this coding system to a sample of speech in some detail, because we believe it will become a valuable tool in many studies of discourse. Using Grimes's system, we outlined each answer and coded each grammatical clause in the answer for its relationship with preceding ones.

Because the previous analysis showed that the difference in liars' effectiveness when talking to active and passive interviewers was constant for transcripts from both replications, we limited our analysis of semantic structure to transcripts from the first replication. The answers to two questions in the active and passive conditions were examined for each of the twelve speakers from the first replication.

Two coders developed an outline of each answer that showed the sequencing of and logical subordination between clauses. For each pair of clauses, if the second elaborated upon or in some way developed a theme introduced by the first, it was considered subordinate to the first. If a clause had an equal weight with or did not develop the idea of the immediately preceding one, its subordination to other preceding clauses was established. Subordination could be nested so that a clause could be subordinate to a preceding one, which in turn could be subordinate to one before it. Up to six levels of subordination occurred. Two clauses were at the same level if they were subordinate to the same dominant clause, or if they were both at the highest level in the outline.

After the two coders discussed and reached agreement on an outline, they independently coded each clause for its relationship to the immediately preceding clause of the same level, and for its relationship to the clause immediately dominating it. Of a sample of 478 relationships coded, two independent judges agreed on 69 percent of them, where the chance level of agreement was 14 percent. Differences were resolved through discussion. We coded the following nine relationships, dropping those of Grimes's categories that did not apply to our corpus.

(1) *Repetition.* The second clause is a restatement or reworking of information presented in the first. Any seemingly new information has already been strongly implied by the first clause (for example, "I was a little paranoid. I thought everyone talked behind my back'').

(2) *Alternative.* The second clause presents an opposing situation, point of view, or argument to that presented in the first (for example, "I don't think she's thinking rationally. If she were thinking rationally . . .''.).

(3) *Conditions.* The second clause places conditions on the material presented in the first. This relationship is similar to that between an "if" or "when" clause in a sentence and the clause that precedes it (for example, "Human life ends when the brain and heart stop functioning").

(4) *Results.* The second clause presents results of the material in the first, implying a cause-and-effect relationship. This relationship is similar to that between "then" clauses and the clauses that precede them in "if-then" and "when-then" constructions (for example, "Recently, because of hanging around people in college, I've been able to get along with other people").

(5) *Summary.* The clause summarizes the preceding information and usually occurs at the end of a section.

(6) *Response.* The content of the second clause is taken largely from the first, but new material is added as well, starting toward a new idea. This relationship is similar to that between a question and a reply, in which the content of the reply is closely related to the question but also includes new information. This category applies only to clauses at the same level (for example, "I think it's up to me. And I think I would consider having an abortion").

(7) *Collection.* The clauses contain completely different and unrelated ideas, although they may be subordinated to the same higher-level clause. Their order is unimportant. This category applies only to clauses at the same level (for example, "She was settled down. She's two years older than I am. She's working").

(8) *Specification.* A subordinate clause provides new details about the information presented in the dominant unit, which could not have been guessed from the dominant unit alone (for example, "I had plans. I was going to get a job").

(9) *Explanation.* A subordinate clause gives a reason for the information presented in the dominant clause, answering the question "why?" (for example, "I did a little research on this. I had a debate on it").

Table 9.2 is an example of a segment adapted from an outlined interview, showing an example of each coding category.

The Structure of Speech to Active
and Passive Interviewers

We examined the influence of listener responsiveness on the structure of speech in two ways. First, we compared the relationships among idea units in the active and passive interviewer conditions to see if a listener's behavior, in general, influenced conversation. Within the active interviewer condition, we also looked at the types of idea units following back channel responses and clauses lacking back channel responses to see how they changed the structure of speech. For each speaker, we computed the percentage of clauses when talking to an active and a passive interviewer that was represented by each coding category. The analyses are based on these means, with the N for each analysis being the number of speakers (N = 12). Table 9.3 shows the comparison of speech to active and passive interviewers.

When comparing speech to active interviewers with that to passive interviewers, we should first note that speakers talked more to active interviewers. They did not introduce more topics (highest-level clauses), but they talked more about each one. The mean number of highest-level clause per answer was essentially the same in the active and passive interviewer conditions (3.87 versus 3.73, $t(11) = .22$, $p > .20$), but when subjects spoke to an active interviewer, they had more subordinate clauses per highest-level clause (6.30 versus 4.61, $t(11) = 3.10$, $p < .001$).

It is interesting to note that when speakers with active interviewers elaborated a major theme, they did not do so by incremen-

TABLE 9.2   The Coding of Semantic Relationships Between Clauses

I. Well, I had a cousin who got pregnant

    A. but with her it was all right (Specification I)

        1. she was settled down (Specification A)

        2. she's older than I (Collection 1; Specification A)

        3. she already has an apartment and car (Collection 2; Specification A)

        4. so everything was settled for her (Summary 3; Summary A)

    B. and she sat down and thought about it (Response A; Result I)

        1. and said "No, I don't want to get married" (Results B)

        2. so she decided not to get married (Repetition 1; Result B)

        3. but she was going to have the kid (Response 2; Results B)

    C. so she had it (Results B; Summary I)

II. And that baby's just the pride of the family (Response I)

    A. 'cause she's really so hyperactive (Explanation II)

    B. and she's really smart (Collection A; Explanation II)

III. But it depends on what the situation is (Collection II)

    A. 'cause I have some friends who have really big plans for the next few years
       (Specification A)

        1. they're going to finish college (Specification A)

        2. then they're going to grad school (Response 1; Specification A)

        3. and I don't think they can do it (Alternative 2; Alternative A)

           a. if they have a child (Condition 3)

        4. and if they have a child (Condition 3; Condition III)

           a. I wonder if they would resent the child's getting in the way of their
              plans (Result 4)

              1. and feel cheated because they can't do what they want
                 (repetition a)

NOTE: The notation in parentheses indicate the relationship of a clause to preceding clauses. If two relationships are shown, the first is to the preceding clause at the same level and the second is to the immediately dominating clause. For example, Summary 4 means that the labeled clause summarizes the preceding material in clause 4. Specification A means that the labeled clause specifies clause A.

tally adding related information. In fact, they were significantly less likely than speakers with passive interviewers to provide specifications of and responses to their own ideas. These are both relationships in which succeeding clauses go beyond an original idea, either by providing elaborating information or by continuing the development of a point.

When talking to an active rather than a passive interviewer, speakers were more likely to paraphrase and repeat an idea and to spell out the consequences and results of previously mentioned ideas. In addition, they elaborated themes by providing counter-

TABLE 9.3 The Semantic Structure of Answers to Active and Passive Interviewers

Relationship	Active Interviewer	Passive Interviewer	t
To preceding clause at same level			
repetition	.06	.06	-.12
alternative	.16	.08	2.86***
response	.24	.36	-1.91*
collection	.40	.31	1.59
condition	.03	.00	2.09*
result	.09	.09	-.66
summary	.02	.07	-1.26
To dominant clause			
repetition	.24	.16	1.80*
alternative	.09	.04	2.47**
specification	.18	.27	-3.87***
condition	.07	.08	.66
result	.19	.15	2.32**
explanation	.18	.19	-.43
summary	.04	.05	-.94

NOTE: These figures represent the proportion of each type of relationship in two answers for each speaker, averaged over twelve speakers.
$*p < .10$; $**p < .05$; $***p < .02$.

themes, or by mentioning limitations on a theme's generality. That is, speakers with active interviewers had a significantly higher proportion of alternatives and conditions in their speech.

These differences between speech to active and passive partners are consistent with our earlier reasoning. It is as if speakers with passive partners went on with their story no matter what. On the other hand, with an active partner they were sensitive to signs that the interviewer did not understand a point, to which they responded by paraphrasing the point to make it clearer or by spelling out the results. In addition, they were sensitive to signs that the interviewer was not convinced by a point, to which they responded by adding alternatives and conditions to meet the interviewer's objections.

*Speech Following or
Not Following Listener Responses*

From the preceding analysis, we know that speakers organized their sequences of ideas differently when talking to active versus passive interviewers, but we do not know precisely how these differences in speech came about. By looking at the presence and absence of back channel responses within the relatively normal conversation to the active interviewers, we can see how speakers altered their sequences of speech. For this exploratory analysis, we made the simplifying assumption that morphologically different back channel responses share a single meaning.

To the extent that a back channel response changes the sequencing of speech, we expected to find different types of relationships between clauses surrounding back channel responses than between clauses not containing them. Table 9.4 shows that this was indeed the case. Within the active interviewer condition, a clause without a back channel response was followed by a subordinate clause about 50 percent of the time, one at the same level about 31 percent of the time, and one at a higher level about 19 percent of the time. After a back channel response, however, speech was less likely to be subordinate and more likely to have moved to a higher level. In addition, speakers were more likely to summarize an argument following a back channel response. Since summaries generally occur at the end of an answer or before changing topic, both results indicate that speakers were more likely to start a change of topic following a back channel response. In addition, speakers were less likely to spell out the results of previously mentioned ideas following a listener response, as if they knew that their partner had drawn the appropriate implications from the facts. At a minimum, these results imply that listener feedback has a direct effect on the organization of speech.

FEEDBACK AND SUMMARIZATION

We were encouraged that listener responsiveness had effects on both conversational outcome and conversational structure. While deception and its detection are common in social interactions, generalization of these results to different conversational situations and outcome measures is problematic. Specifically, the deception situation is atypical, in that speakers and listeners have conflicting goals—to lie and to detect deception, respectively. This leads to unsatisfyingly ambiguous predictions about whom conversational coordination should benefit most, the deceiver or the detector of deception. The present research attempts to overcome these limitations by providing a clear demonstration that listener responsiveness increases conversational effectiveness, and by ex-

TABLE 9.4 Clauses Following and Not Following Listener Responses

	Following	Not Following	t
Level of next clause			
higher	.31	.19	2.00*
same	.32	.31	.12
lower	.38	.50	−2.33**
Relationship to preceding clause at same level			
repetition	.05	.06	−.29
alternative	.16	.14	.27
response	.25	.31	−.50
collection	.39	.37	.21
condition	.06	.01	1.62
result	.06	.11	−1.25
summary	.03	.00	1.87*
Relationship to dominate clause			
repetition	.27	.23	.83
alternative	.07	.10	−.52
specification	.21	.14	1.13
condition	.08	.06	.38
result	.11	.25	−3.26***
explanation	.21	.18	.69
summary	.06	.04	.81

NOTE: These figures represent the proportion of each type of relationship in two answers for each speaker, averaged over twelve speakers.
*p < .10; **p < .05; ***p < .02.

amining several processes as a result of which such an increase could occur. We studied the role of listener responsiveness in conversational effectiveness by having speakers summarize a movie to one or two listeners.

Feedback may provide speakers with two somewhat different types of information. First, feedback may provide information about the reaction of a generalized audience. To the extent that feedback is obligatory—that it, that any member of the language community would have provided substantially the same feedback at substantially the same places in the stream of speech—feedback from any one listener will provide a speaker with information about how the language community as a whole understands his or her speech. In this case, feedback from any listener should improve the speaker's performance for all listeners. Second, feedback may provide information about the reactions of a particular listener. To the extent that feedback is optional—that is, with different listeners giving different feedback at different times—it can provide a speaker with information about a particular listener's unique understanding, agreement, and needs. In this case we would expect that the person giving the feedback understands the speech better than any other person overhearing the same conversation.

To examine this possibility, we had speakers summarize a movie to pairs of listeners. In each pair, one of the listeners, called the "participant," gave feedback that could influence the speaker. The other listener, called the "eavesdropper," could not influence the speaker, either because the eavesdropper was instructed not to give feedback or because the feedback never reached the speaker. If the feedback was providing information about the generalized audience, both listeners were expected to understand the movie better as the speaker got more feedback from one of them. On the other hand, if the feedback was providing information about a particular listener's reaction, the participant was expected to understand the movie better than the eavesdroppers, especially when the participant provided a normal amount of feedback. Of course, feedback can serve both functions.

To show that feedback can coordinate an interaction between speaker and listener, we also need to demonstrate (1) that listeners actually signal what they know via feedback (that is, that listener feedback changes along with changes in listeners' knowledge and understanding); and (2) that speakers are sensitive to these signals (that is, that their speech also changes along with changes in listeners' knowledge and understanding). To conduct this test, we selected a number of scenes from the movie that was to be summarized to listeners, manipulating listeners' knowledge about the scenes prior to the summary. For some scenes, listeners saw excerpts from the movie to be described that should have made them knowledgeable about these scenes. For other scenes, they saw excerpts from a different but related movie, which should have made them confused when parallel scenes were described in the summary. And for still other scenes, they saw no excerpts from either movie.

We measured the amount and type of feedback listeners gave in response to descriptions of scenes for which they were differentially informed. We also measured the completeness, accuracy, and efficiency of the speakers' descriptions of scenes about which listeners had been differentially informed. Although we made no specific prediction, we expected, for example, that speakers would give shorter descriptions of scenes about which listeners had been informed, and that listeners would ask more questions about scenes about which they were confused (see Kraut, Lewis, & Swezey, 1982, for a fuller discussion of this research).

Methods

We showed 78 speaker subjects a cowboy movie and then had them summarize it to one or two listeners. The movie *Bend of the River* is a ninety-minute western in which an ex-outlaw helps some pioneers get and retain supplies so that they can establish a settlement. Speakers and listeners were in separate rooms, and all communication occurred over an audio link.

We experimentally varied three independent variables. First we manipulated the amount of feedback the speaker got from the listener. This manupulation involved three levels, unrestricted or free feedback, restricted or back channel only feedback, and no feedback at all.

Second, we manipulated the listener's participation status—which of two listeners, when two were present, provided the feedback to the speaker and which was the eavesdropper. In the unrestricted feedback condition, the eavesdropper was told simply to listen to the speaker's summary and to provide no feedback at all. In the limited feedback condition, both listeners had identical instructions. However, one listener's microphone was disconnected after thirty seconds of the summary. Thus, in each group with an active listener and an eavesdropper, the two listeners heard exactly the same description of the movie, but the description was influenced by only one of them.

Finally, we manipulated listeners' prior knowledge—the amount they knew about seventeen preselected scenes before hearing the speakers' summaries. Some listeners saw a combination of film clips designed to make them knowledgeable about some scenes, confused about other scenes, and ignorant about still other scenes, while others saw no previews at all.

After the speakers gave a ten-minute summary of the movie, both speakers and listeners completed dependent measures about the movie and the conversation. These included the following:

(1) *Quality of the speakers' summary.* We transcribed the speakers' summaries and then measured several characteristics that reflected their quality. Two investigators coded each speaker's summary for the completeness and accuracy with which it covered each of 107

elements of theme, plot, and characterization important to understanding the movie.

(2) *Summary length.* As a measure of summary length, we counted the number of words in the speakers' summaries.

(3) *Listener comprehension.* Immediately after hearing the speaker's summary, the listeners themselves summarized the movie. Their recorded descriptions of the seventeen preview scenes were judged by two independent coders for overall quality and for completeness and accuracy. Listeners then answered a fact test composed of thirty multiple choice questions about specific plot and character details. They next rated the two main characters in the film on thirty evaluative adjectives. Speakers also had rated the main characters on these scales. We were interested in the similarity of these evaluations. For each speaker-listener pair, we produced two indexes of similarity by computing the Pearson correlation between the speakers' and listeners' descriptions of the hero and the correlation between their descriptions of the villain. We also computed an overall comprehension measure by standardizing these six listener comprehension measures and taking their mean.

(4) *Listener responses.* We also attempted a preliminary classification of the listener responses during the summaries. They will be discussed below.

(5) *Assessments of the conversation.* After the summaries, listeners judged the conversations. They rated the clarity of the speakers' summaries and the extent to which a speaker had overcome the listener's lack of understanding. In addition, both assessed their satisfaction with the conversation, using a scale developed by Hecht (1978). This scale was used as a covariate in some of the analyses that follow to determine the extent to which feedback per se, independent of participants' satisfaction with a conversation, can change the quality of an interaction.

Effects of Feedback on Listener Comprehension

One of our major concerns was the way in which the amount of feedback speakers received during their summaries influenced the quality of their summaries and listeners' comprehension of them. In general, the results showed that as speakers received more feedback from a partner, listeners understood their summaries better and were more influenced by them. This effect occurred for both participant-listeners' and eavesdropper-listeners' comprehension of the summary (see Tables 9.5 and 9.6).

As the tables show, listeners correctly answered more questions on the fact test when their speakers received more feedback. Moreover, listeners' descriptions of the hero and villain were more

TABLE 9.5 Effects of Feedback to a Speaker on Participants' Comprehension (N = 57)

Feedback to Speaker	Dependent Variables								Covariates		
	Fact Test (%)	Hero Similarity	Villain Similarity	Summary Completeness	Summary Accuracy	Global Quality	Overall Comprehension	Judged Clarity	Length	Idea Units	Listeners' Satisfaction
Free	62.0	.731	.526	.565	1.58	3.18	.384	5.26	27.4	119	5.14
Limited, no preview	54.3	.633	.315	.407	1.51	2.39	-1.65	3.76	21.4	119	3.62
Limited, preview	58.5	.655	.449	.546	1.58	2.97	.231	5.32	24.4	123	4.57
Correlation amount of feedback	.20	.19	.21	.17	.04	.21	.26**	.24*	.23*	-.03	.40***
partialling length	.19	.21	.16	.09	.03	.15	.21	.18			
partialling idea units	.21	.19	.22*	.18	.05	.22*	.27**	.25*			
partialling satisfaction	.10	.10	.11	.10	.02	.09	.12	-.01			

NOTE: Correlations are with feedback to speaker as an experimental variable, with free coded 1 and limited feedback coded -1.
*p < .10; **p < .05; ***p < .02.

249

TABLE 9.6 Effects of Feedback to a Speaker on Eavesdroppers' Comprehension (N = 78)

Feedback to Speaker	Fact Test (%)	Hero Similarity	Villain Similarity	Summary Completeness	Summary Accuracy	Global Quality	Overall Comprehension	Judged Clarity	Length	Idea Units	Listeners' Satisfaction
				Dependent Variables					Covariates		
Free	60.7	.716	.585	.471	1.66	2.71	.303	5.12	27.4	119	4.53
Limited, no preview	51.7	.679	.435	.374	1.51	2.37	-.122	3.53	21.4	119	3.64
Limited, preview	54.3	.633	.315	.407	1.51	2.39	-.165	3.76	21.4	119	3.62
No feedback	52.7	.527	.435	.353	1.42	2.19	-.295	4.36	18.1	113	4.28
Correlation with amount of feedback	.21*	.26***	.20*	.18	.24**	.17	.31***	.17	.34	.09	.06
partialling length	.17	.21*	.14	.08	.17	.12	.22*	.18			
partialling idea units	.20*	.26***	.19*	.18	.24**	.16	.30***	.19*			
partialling satisfaction	.21*	.25**	.19*	.17	.24**	.16	.29***	.14			

NOTE: Correlations are with feedback to speaker as an experimental variable, with free coded 1, limited feedback coded 0, and no feedback coded −1.

$*p < .10; **p < .05; ***p < .02.$

similar to the speakers' descriptions when their speakers received more feedback. Examination of the listeners' summaries also shows better understanding with more feedback.

An alternative explanation: effects of the length and quality of the speaker' summaries. We believe that feedback allowed listeners to understand the summaries better because it helped to coordinate the summaries with the listeners' needs. Some data support this interpretation. For example, the listeners believed that the speakers' summaries were clear when the speakers received more feedback. An alternative explanation, however, is that with more feedback, speakers simply talked more and provided more information about the movie (see Matarazzo, 1964). Indeed, with more feedback speakers gave longer summaries, but they provided only slightly more idea units in them. Moreover, the effects on listeners' comprehension were only slightly reduced when the length of the summaries and the number of idea units in them were controlled via partial correlations (see Tables 9.5 and 9.6).

Effects of Participation Status

Another of our major hypotheses was that to the extent that feedback individuates conversation—that is, tailors it to a listener giving feedback—participant listeners would understand summaries better than eavesdropper-listeners, even though they were listening to identical summaries. A test of this hypothesis compared the comprehension measures for participants and eavesdroppers, dropping the no-feedback cell in which there were no participant listeners. Table 9.7 shows the relevant data in support of this hypothesis; participants' summaries were indeed better than eavesdroppers'. Coders judged them to be of higher quality, and they were more complete. In addition, participants had higher scores on their fact tests and judged the speakers' summaries to be clearer.

Perhaps more surprising is the finding that variations in the quality of the speakers' summaries influenced the participant listeners more than they influenced the eavesdroppers. In general, when a speaker gave a better summary, the participants' summaries were also better, but the eavesdroppers' summaries were not. Table 9.8 shows the correlation of the number of idea units in the speakers' summaries with various measures of listener comprehension for the matched participants and eavesdroppers. For example, when speakers included more idea units in their sum-

TABLE 9.7 Effects of Participant Status on Listener Comprehension (N = 57)

Participant Status		Dependent Variables							
	Fact Test (%)	Hero Similarity	Villain Similarity	Completeness of Summary	Accuracy of Summary	Global Quality	Overall Comprehension	Judged Clarity	Listeners' Satisfaction
Eavesdropper	54.5	.658	.461	.404	1.53	2.45	-.04	4.39	4.06
Participant	58.2	.673	.447	.506	1.56	2.85	.15	4.78	4.44
Correlation participant status	.22	.05	-.05	.33***	.07	.28**	.23*	.26**	.24*
partialling satisfaction	.17	.00	-.10	.32***	.06	.24*	.18	.16	

NOTE: Correlations show whether participants understood the movie better than did eavesdroppers. They are the correlation-transformation of the single sample t-test on differences $(r = t/[(N - 2 + t**2] **.5)$.

*$p < .10$; **$p < .05$; ***$p < .02$.

TABLE 9.8 Effects of Quality of Speakers' Summaries on Participants' and Eavesdroppers' Comprehension (N = 57)

Participant Status		Dependent Variables						
	Fact Test	Hero Similarity	Villain Similarity	Completeness of Summary	Accuracy of Summary	Global Quality	Overall Comprehension	Judged Clarity
Eavesdropper	.05	-.17	.21	-.05	-.11	.06	.08	.02
Participant	.23*	.01	.19	.32***	.25	.46***	.38***	.32***
Correlation participant status	.15	.13	-.01	.30**	.28**	.32***	.28**	.28**

NOTE: Entries are correlations of the dependent variables with the number of idea units in the speakers' summaries. The bottom row shows whether the association is stronger for participants or eavedroppers and is the correlation transformation of the single sample t-test (see Table 9.4).

*p < .10; **p < .05; ***p < .02.

maries, the summaries from participant listeners were of higher quality, while the eavesdroppers' were not. This same pattern occurred for all but one of the measures of listener comprehension, and significantly so for the overall measure of comprehension, for the accuracy and completeness of listeners' summaries, and for their assessment of the speakers' clarity.

Effects of Prior Knowledge on Listener Responses and Summarization

To demonstrate that listener feedback actually coordinates conversation, we need to examine the process of coordination as well as the outcomes we have presented so far. Specifically, we need to show that listener responses reflect differences in their knowledge or understanding, and that speakers are sensitive to these differences.

We analyzed the number of listener responses from each participant who viewed a preview-tape using an amount of feedback (full versus restricted)-by-scene type (old, new, or confusing) analysis of variance, with scene type as a within-subject factor.

The results for scene type show that listeners made more substantive comments when speakers were summarizing scenes for which the listeners had seen previews, either old or confusing. Thus they asked more multiword questions and added more new information on their own. In addition, listeners provided somewhat more new information about old scenes than about scenes for which they had seen confusing previews. The interaction of amount of feedback and scene type for the new information measure shows that the difference between previewed and nonpreviewed scenes occurred only in the unrestricted feedback condition.

The amount of listeners' foreknowledge had an influence on certain characteristics of the speakers' summaries as well. In general, the results showed that speakers spoke more efficiently about scenes for which listeners had seen any preview, providing more information in fewer words. Here, the analysis was an amount of feedback (full, restricted or none)-by-scene type (old, new, or confusing) analysis of variance.

Speakers described scenes for which listeners had seen previews more completely. In addition, they used fewer words to describe these scenes. The interaction of amount of feedback and scene type also affected length: Speakers summarized preview scenes most compactly when listeners were allowed to give full feedback. On the other hand, the accuracy of speakers' summaries were unaffected by either the amount of feedback listeners provided or their prior knowledge of scenes.

Summary

In summary, these data provide evidence that speakers can use listener feedback to tailor what they say to what a listener needs to know. Listeners give feedback that indicates how informed they are about the topic at hand, and speakers modify what they have to say on the basis of that feedback.

DISCUSSION

This research has demonstrated experimentally that feedback influences conversational outcomes. It has done so by examining several of the mechanisms through which feedback operates. For example, we have shown that listener feedback can influence how well a speaker lies. When speakers talked to active rather than passive interviewers, judges reading transcripts of these interviews had an easier time identifying lies and truths. Listener responsiveness seemed to have an effect in part by providing speakers with feedback that coordinated their speech to what judges needed to know. In this mixed-motive situation, the resulting clarity of speech seemed to help detectors of deception more than liars.

In addition, we have shown that in terms of information transfer, communication was better—that is, the more feedback the communicator got, the more all listeners benefited. We have also shown that feedback seemed to individuate speech; the person giving the feedback benefited more than another person who overheard the same conversation but had no opportunity to influence it. Participants understood the communication better and were more influenced by variations in its quality.

One of the mechanisms through which feedback was operating was to show the speaker what a listener already knew, thus allowing the speaker to modify his or her speech in response to the listener's needs. For example, in the deception experiment the presence of some responses seemed to signal the speaker that the listener understood and agreed with a point sufficiently, and that the speaker could get on with the story. In the passive interviews, when speakers got little feedback from their partners, they continued with the details of their answer, not knowing when to backtrack for repetition, alteration, or further explanation. On the other hand, in the context of a normally responsive listener, a lack of feedback probably indicated the listener's lack of understanding. In any case, speakers tended to finish an idea and move on with their story after they received a head nod or a verbal signal from their partner.

In the cowboy movie experiment, we demonstrated that listeners show what they know through their use of substantive questions and comments. In addition, speakers, who were initially unaware of the listeners' prior knowledge, talked more efficiently about those topics on which the listeners were knowledgeable. Thus it appears that, among other functions, feedback can regulate the informational density of speakers' communication.

Although in both studies we showed that speakers talked more to responsive listeners, the details are inconsistent with a reinforcement model (Matarazzo, Wiens, Sastow, Allen, & Weitman, 1964). The presence of a back channel response decreased a speaker's tendency to persevere on a point.

As we stressed in the introduction, feedback in general (and back channel feedback in particular) is but one of the mechanisms speakers use to modify what they have to say and to adapt it to a particular audience. At this point we will discuss more generally the way in which people organize speech and adapt it to an audience in order to examine the role of feedback in this overall process.

The first point to recognize is that people talk with social motives in mind—for example, to exchange information, give orders, persuade, impress, express feelings, and the like. In order to achieve these ends with a particular audience, speakers must plan their speech with that audience in mind. Since speech is hierarchically organized, its planning can be accomplished at several levels. For example, in narrating a story plot, speakers can shape the organization of the story they are telling, the structure of particular sentences, and the selection of particular vocabulary to fill slots in the sentence frames (see Chafe, 1977). Certainly a college freshman retelling *Alice in Wonderland* to an English professor with a Freudian orientation, or at home to a younger sister, would make changes on all these levels. The planning at different levels occurs at different intervals before actual speech production. Some parts, like the main points to be covered, may be planned relatively far in advance, while other parts, like word choice, may be planned just before speech production.

After they have spoken, people have two mechanisms available to them that can aid in evaluating what they have just said. They can compare what they wanted to say with the interpretation a typical listener might make by listening to their own speech, both from the standpoint of what Mead (1934) has called the generalized other, and from the standpoint of a particular audience. In many ways this is no different from planning one's speech with different audiences in mind.

In addition to this totally internal evaluative process, speakers can pay attention to and interpret explicit feedback from their audience. It is likely that self-evaluation is the basic adaptive mechanism here, which explicit feedback merely supplements. On the basis of the outcome of either type of feedback, a speaker can decide to continue with the preplanned speech or to backtrack and modify what was said at one of many levels of planning.

Now let us consider the case of explicit audience feedback in more detail. To the extent that audience feedback causes speakers to modify what they would have said, it can have one of two consequences. First, it could change their social goals for a conversational episode. For example, if listeners somehow signal to speakers that they are bored, speakers will often shift their conversational goal from telling a clear story to entertaining the listeners, or to terminating the conversation with them. Similarly, if listeners begin asking difficult and/or hostile questions, speakers are likely to shift their goal from informing the listeners to placating them.

Alternately, feedback may cause speakers to modify what they were saying in order to better achieve their original goals. That is, they may back up to the start of a convenient planning unit and modify what they have said or plan to say in order to make their speech clearer, more persuasive, or more interesting.

Let us consider the circumstances under which we would expect speakers to be most influenced by or to make the most use of explicit, external feedback. First and obviously, for feedback to influence a speaker, the listener must give adequate and comprehensible feedback. When the listener is deadpan, the speaker by necessity adopts different methods to determine whether his or her speech is comprehensible to the listener.

Another condition for speakers' effective use of feedback is that the speaker must be motivated to pay attention and respond to it. This should occur most to the extent that speakers are trying to get their views across to a particular listener in spontaneous rather than pre-planned conversation. Thus, speakers may ignore particular feedback if they believe it to be unrepresentative of what a general audience would understand or if they are committed to their preplanned text.

Even in cases where speakers receive adequate feedback to which they are motivated to respond, the feedback will individuate the speech more—that is, make it better suited to the person providing the feedback—if several conditions are met. The more a speaker knows about the perspective of and information available to a listener, the more he or she can take this knowledge into account in phrasing material so that the listener will understand it. Specifically, speakers can refer to facts known to both themselves and a listener in a shorthand form that is likely to confuse an eavesdropper. For example, they can rely on definite articles and pronouns more if they are assured that listeners know to what they are referring. A speaker can gain this knowledge either from relying on feedback from the partner or from knowing the prior history and perspective of the partner. Thus a speaker can be assured that the sentence, "He shouldn't have done it," has been understood either if the listener gives a slow, sympathetic nod afterward, or if the speaker knows from previous conversation or history that the listener knows about the act and the person mentioned.

In summary, the maximum individuating effects of feedback should come in spontaneous conversation, when a speaker is very familiar with his or her partner prior to speaking, when the partner provides a lot of clear feedback, and when the partner and the eavesdropper differ widely in their initial knowledge. Given these conditions, it is surprising that in the present experiments listener responsiveness and listener participation status had reliable effects on conversational outcome and process. For example, in the cowboy movie experiment, feedback influenced conversational process and outcome even in a constrained setting in which the conversations were not spontaneous, the feedback was minimal, the speakers and listeners shared no history, and the participant listeners and eavesdroppers were similar. These influences are likely to be much stronger in more interactive, natural conversation.

We have not meant to imply here that explicit feedback from a partner is the only way that a speaker can adapt what he or she has to say to a listener's informational needs, that speakers invariably respond to feedback, or that feedback serves only to regulate the informational density of speech. We have argued that speakers have other mechanisms available. Even if feedback is a major mechanism in transforming speech, speakers may fail to respond to many explicit feedback signals from their partners, either because the feedback conflicts with their self-evaluation or because they have not come to the end of a convenient planning unit.

In addition to whatever functions feedback serves in regulating the semantic content of speech, it has other conversational and interpersonal functions as well. By showing the listener's interest, it demonstrates a continuing commitment on the listener's part to engage in the conversation. Bored looks and a lack of feedback cause at least socially sensitive speakers to stop talking. In addition to the information that feedback carries about the conversational process per se, it is often interpreted by participants in a conversation as indicating the state of a relationship between two parties. Smiling and head nodding are signs of affiliation, as well as understanding and interest (see Kraut & Johnston, 1979).

REFERENCES

Bolinger, D. (1973). Truth is a linguistic question. *Language, 49,* 539-550.

Chafe, W. (1977). Creativity in verbalization and its implication for the nature of stored knowledge. In R. Freedle (Ed.), *Discourse production and comprehension.* Hillsdale, NJ: Lawrence Erlbaum.

Clark, H. (in press). Lanuage use and language users. In G. Lindzey & E. Aronson (Eds.), *Handbook of social psychology (3rd ed.).* Reading, MA: Addison-Wesley.

Clark, H., & Clark, E. (1977). *Psychology and language.* New York: Harcourt Brace Jovanovich.

Cronbach, L. (1955). Processes affecting scores on "understanding of others" and assumed similarity. *Psychological Bulletin, 52,* 117-193.

Dittmann, A., & Llewellyn, L. (1968). Relationship between vocalizations and head nods as listener responses. *Journal of Personality and Social Psychology, 9,* 79-84.

Duncan, S. (1974). On the structure of speaker-auditor interaction during speaking turns. *Language in Society, 2,* 161-180.

Grice, H. (1975). Logic and conversation. In. P. Cole & J. Morgan (Eds.), *Syntax and semantics: Vol. 3. Speech acts.* New York: Academic Press.

Grimes, J. (1975). *The thread of discourse.* The Hague: Mouton.

Hale, C. L. (1980). Cognitive complexity-simplicity as a determinant of communication effectiveness. *Communication Monographs, 47,* 304-311.

Hale, C. L., & Delia, J. G. (1976). Cognitive complexity and social perspective-taking. *Communication Monographs, 43,* 195-203.

Hecht, M. (1978). The conceptualization and measurement of interpersonal communication satisfaction. *Human Communication Research, 4,* 253-264.

Hemsley, G. (1977). *Experimental studies in the behavioral indicants of deception.* Unpublished doctoral dissertation, University of Toronto.

Kendon, A. (1967). Some functions of gaze-direction in social interaction. *Acta Psychologica, 26,* 22-63.

Kent, G., Davis, J., & Shapiro, D. (1978). Resources required in the construction and reconstruction of conversations. *Journal of Personality and Social Psychology, 36 ,* 13-22.

Krauss, R., & Bricker, P. (1967). Effects of transmission delay and access delay on the efficiency of communication. *Journal of the Acoustical Society of America, 41,* 286-292.

Krauss, R., Garlock, C., Bricker, P., & McMahon, L. (1977). The role of audible and visible back-channel responses in interpersonal communication. *Journal of Personality and Social Psychology, 35,* 523-529.

Krauss, R., & Weinheimer, S. (1964). Changes in the length of reference phrases as a function of social interaction: A preliminary study. *Psychonomic Science, 1,* 113-114.

Krauss, R., & Weinheimer, S. (1966). Concurrent feedback, confirmation, and the encoding of referents in verbal interaction. *Journal of Personality and Social Psychology, 4,* 342-346.

Kraut, R. (1978). Verbal and nonverbal cues in the perception of lying. *Journal of Personality and Social Psychology, 36,* 380-391.

Kraut, R. (1980). Humans as lie detectors: Some second thoughts. *Journal of Communication, 8,* 209-216.

Kraut, R., & Poe, D. (1981). Behavioral roots of person perception: Deception judgments of customs inspectors and laymen. *Journal of Personality and Social Psychology, 39,* 784-798.

Kraut, R. E., & Higgins, E. T. (in press). Communication and social cognition. In R. Wyer & T. Scrull (Eds.), *Handbook of social cognition.* Hillsdale, NJ: Lawrence Erlbaum.

Kraut, R. E., & Johnston, R. (1979). Social and emotional messages of smiling: An ethological approach. *Journal of Personality and Social Psychology, 37,* 1539-1553.

Kraut, R. E., Lewis, S. H., & Swezey, L. (1982). Listener responsiveness and the coordination of conversation. *Journal of Personality and Social Psychology, 43,* 718-731.

Labov, W., & Fanshel, D. (1977). *Therapeutic discourse.* New York: Academic Press.

Leathers, D. (1979). The information potential of the nonverbal and verbal components of feedback responses. *Southern Speech Communication Journal, 44,* 331-354.

Matarazzo, J., Wiens, A., Sastow, G., Allen, B., & Weitman, M. (1964). Interviewer mm-hmm and interviewee speech duration. *Psychotherapy: Theory, Research, and Practice, 1,* 109-114.

McGuire, W. J. (1969). Tha nature of attitudes and attitude change. In G. Lindzey & E. Aronson (Eds.), *Handbook of social psychology* (2nd ed.). Reading, MA: Addison-Wesley.

Mead, G. (1934). *Mind, self, and society.* Chicago: University of Chicago Press.

Rosenfeld, H. (1972). The experimental analysis of interpersonal influence processes. *Journal of Communication, 22,* 424-442.

Rosenfeld, H. (1978). Conversational control functions of nonverbal behavior. In A. Siegman & S. Feldstein (Eds.), *Nonverbal behavior and communication.* Hillsdale, NJ: Lawrence Erlbaum.

Rosenthal, R., Hall, J., DiMatteo, M., Rogers, P., & Archer, D. (1979). *Sensitivity to nonverbal communication: The PONS test.* Baltimore, MD: Johns Hopkins University Press.

Yngve, V. (1970). On getting a word in edgewise. In *Papers from the sixth regional meeting, Chicago Linguistics Circle.* Chicago: Chicago Linguistics Circle.

Zuckerman, M., DePaulo, B., & Rosenthal, R. (1981). Verbal and nonverbal communication of deception. In L. Berkowitz (Ed.), *Advances in experimental social psychology: Vol. 14.* New York: Academic Press.

PART IV

SOCIAL COGNITIVE AND STRATEGIC PROCESSES IN ADULTS: THE NEGOTIATION OF RELATIONSHIPS

10

The Evolution of Impressions in Small Working Groups:
Effects of Construct Differentiation

Barbara J. O'Keefe

Most views of impression formation take it to be a process of drawing dispositional inferences from observations of behavior and organizing those inferences as a general impression of the target figure's personality. The cognitive mechanism that outputs inferences about personality is generally described as an "implicit theory of personality" (Bruner & Taguiri, 1954); that is, an organized set of beliefs about what dispositions people have, what behaviors are symptomatic of each disposition, and which dispositions generally co-occur in the character of a target.

There are a variety of specific models of the structure and functioning of implicit theories of personality, but one common view of these cognitive structures is grounded in Kelly's (1955) personal construct theory. According to Kelly, an individual's cognitive system comprises a number of elements called constructs. Constructs are bipolar dimensions of judgment organized hierarchically into subsystems according to the range of application of constructs contained in the subsystem. Thus they form construct *domains* that are used in representing events and objects from particular realms of experience. The construct subsystem used in construing people and their behavior is commonly referred to as the "interpersonal domain." Implicit theories of personality are thus frequently conceptualized as one of the most important subsets of the set of constructs contained in the interpersonal domain.

In elaborating this analysis of social cognition, Crockett (1965) argued that the development of construct systems follows Werner's (1957) orthogenetic principle: "Whenever development occurs, it proceeds from a state of relative globality and lack of differentiation to a state of increasing differentiation, articulation, and hierarchic integration" (p. 126). In applying this principle to the analysis of construct system development, Crockett (1965) and Delia (1976) have argued that as construct systems develop, the interpersonal domain becomes more differentiated and articulated (constructs increase in number and become better adapted to the task of construing the psychological processes of others), and more integrated (both hierarchically, through the development of abstract constructs that subsume and organize more concrete constructs, and implicationally, through the development of inferential connections among constructs). Crockett and Delia reasoned that if the interpersonal construct system develops in this way, then as people mature the impressions they form using this system should change in regular ways. Specifically, as the age of the perceiver increases, his or her impressions should become more abstract (focusing more on internal psychological states and less on concrete, observable features of appearance and behavior), more integrated, and more differentiated. A number of investigations have shown that as age increases, a person's free-response descriptions of others change in just these ways (see, for example, Biskin & Crano, 1977; Delia, Burleson, & Kline, in press; Scarlett, Press, & Crockett, 1977).

This general analysis of construct system development has also been extended to the analysis of individual differences among adult perceivers. Within any age-homogenous adult population, some individuals have more developmentally advanced construct systems (producing relatively differentiated, abstract, and integrated impressions of others) and some individuals have relatively impoverished construct systems (producing less abstract, integrated, and differentiated impressions).

The application of this analysis of construct system development to research on impression formation and its role in interpersonal behavior has produced two sets of findings. One body of findings suggests that as interpersonal construct systems develop, the ways in which people form impressions change in a number of important ways. As their impressions become more abstract, integrated, and differentiated, people also show increasing abilities to transcend an evaluative orientation in judging others, are less reliant on simple evaluative consistency rules in integrating information, and are less influenced by immediate context in making attributions and forming attitudes toward others. A second body of findings suggests that as interpersonal construct systems develop, the manner in which people pursue interpersonal tasks changes. Their behavior is less dominated by evaluative consistency rules, is more likely to display recognition of the differing wants and perspectives of others, and is more likely to reflect attention to subsidiary interactional aims such as identity management (for recent reviews, see O'Keefe & Sypher, 1981; O'Keefe & Delia, 1982, in press).

However, the research that supports these generalizations, like most research on social cognition, has abstracted the operation of interpersonal construct systems from their normal interactional contexts of use. Investigations of the role of construct system development in impression formation have focused primarily on the formation of first impressions and generally have required subjects to form impressions of fictitious targets from provided trait information. Investigations of the role of construct system developments in influencing interpersonal behavior have generally relied on the use of role-played message construction tasks and self-report questionnaires, avoiding the issue of how differences in behavior associated with construct system development might make a difference in actual interaction contexts.

This chapter reports the results of an initial investigation of the role of interpersonal construct system developments in influencing impressions and behavior under more naturalistic conditions. The initial investigation was designed to provide information about the influence of perceivers' and targets' levels of interpersonal construct differentiation on the evolution of impressions over the course of developing acquaintance. The design of the study was simple: Undergraduate students who were initially unacquainted were assigned to work together in small groups on a graded project. They were asked to report their impressions of each of their fellow group members at regular intervals throughout the course of the project. The analysis of these impressions focused on the respective roles of perceiver and target construct differentiation in influencing the structure and content of impressions and their pattern of evolution.

The chapter is organized into three sections. The first section discusses research related to the question of how perceiver and target construct differentiation influence the structure and content of impressions. The second section describes the project and its results. The third section discusses the implications of the findings for models of the impression formation process.

CONSTRUCT DIFFERENTIATION AND IMPRESSION FORMATION

A person's level of interpersonal construct differentiation has been shown to be related to (a) the manner in which a person acquires and integrates information about others and (b) the variability and complexity of behavior that person produces in pursuing interpersonal tasks. Thus, a person's level of construct differentiation should influence the structure and content both of impressions that he or she forms and of impressions formed of that person. In this section I summarize research relevant to this hypothesis.

Construct Differentiation as a Perceiver Characteristic: Impact on Impressions

Construct differentiation and impression structure. Much of the research by Crockett, Delia, and their coworkers has focused on the analysis of structural characteristics of impressions. Two features have been of particular interest: differentiation and organization.

Impression differentiation refers to the number of separable elements (constructs) employed in a free-response description. Impression differentiation has generally been interpreted as reflecting the number of judgmental dimensions a perceiver has employed in construing a target. A very consistent body of findings suggests that relative differentiation is a stable and characteristic feature of an individual's developing style in impression formation. In general, a person who (relative to some comparison group) has an elaborated impression of one person will have elaborated impressions of other targets as well; similarly, a person who reports an impoverished impression of one person has generally impoverished impressions. In fact, the tendency to use a consistent style in elaborating impressions is strong enough that the two-peer version of Crockett's (1965) Role Category Questinnaire has been adopted as a standard measure of construct differentiation rather than the original eight-role version (for a review of relevant research, see O'Keefe & Sypher, 1981).

The Role Category Questionnaire (RCQ) asks a subject to identify a set of target figures the subject knows well and who fit a set of specified roles (for example, a liked and a disliked peer in the two-peer version) and to describe each of these target figures. The number of separable elements or constructs contained in each impression is counted, and the total number of constructs produced in responding to the questionnaire is determined by summing across roles. This produces a single index of impression differentiation, which is interpreted as reflecting the number of judgmental dimensions a perceiver normally has available in construing others (construct differentiation). Differentiation scores on the RCQ are essentially independent of such contaminating influences as general intelligence or verbal fluency (O'Keefe & Sypher, 1981; O'Keefe & Delia, 1982) and are good predictors of differentiation scores on any other free-response impression formation task.

Impression organization refers to the degree to which variability and/or inconsistency in the behavior of a target is recognized and reconciled in an integrated impression. Elaborating Gollin's (1954-1955) classic observation that people seem to employ one of several distinct strategies for integrating inconsistent information, Crockett, Press, Delia, and Kenny (1975) developed a system for describing the degree of organization manifest in a free-response impression. The first version of this scoring system classified impressions into one of five main categories depending on the degree to which impressions reflected some recognition of evaluative inconsistency in information provided about a target and to which they made some effort to reconcile and explain the inconsistency. (This system was designed for use in investigations such as those employing variants of Asch's (1946) experimental paradigm.) A second version of this scoring system, adapted for use in analyzing naturally formed impressions of real people, was subsequently developed. The second system employs categories parallel to those in the earlier system but which are defined and ordered by the degree to which the perceiver recognizes, represents, and reconciles variability in the behavior of a target in forming an impression.

Like impression differentiation, the level of organization of a perceiver's free-response impression is related to his or her RCQ construct differentiation score. A series of investigations has shown that construct differentiation is generally related to the level of organization of a perceiver's impressions, whether these are created experimentally from provided information (Crockett, Gonyea, & Delia, 1970; Nidorf & Crockett, 1965; Rosenkrantz & Crockett, 1965; O'Keefe, Delia, & O'Keefe, 1977), or naturally from observations of behavior (Delia, Clark, & Switzer, 1974). However, this relationship has been attenuated experimentally through varying the similarity of target to perceiver and through varying instructional set (Meltzer, Crockett, & Rosenkrantz, 1966; Delia, 1972; Crockett, Mahood, & Press, 1975; Press, Crockett, & Delia, 1975). O'Keefe and Sypher (1981) note that the association of construct differentiation and level of organization in impressions is often quite weak. Nevertheless, within age-homogenous adult populations, construct differentiation is generally significantly related to impression organization, and across the age span from early childhood to adulthood, as construct differentiation increases, so does impression organization.

Construct differentiation and impression content. As construct differentiation increases, the judgmental dimensions a perceiver employs in construing others become more abstract and comprehensive, with the consequence that the content of a perceiver's impressions becomes less dominated by immediate goals and less tied to overtly observable features of appearance and behavior.

Construct abstractness refers to the degree to which a person's beliefs about others reflect a focus on internal psychological states, as opposed to physical appearance and specific behaviors. Construct differentiation and abstractness are developmentally linked; as a person's impressions become more differentiated, they also become more abstract. This relationship is usually significant even within age-homogenous groups of adults and children and across the age span with the effects of age partialed out (Applegate, Kline, & Delia, 1980; Burke, 1979; Delia, Kline, Burleson, Clark, Applegate, & Burke, 1980).

Construct comprehensiveness refers to the degree to which an individual's constructs are used with a restricted or broad range of application. Noncomprehensive constructs are used in assessments of only a limited range of targets (for example, those who occupy some specific social role), whereas comprehensive constructs are used in judging a wide range of targets. Construct differentiation and comprehensiveness have been found to be related features of interpersonal construct systems within adult and young adult populations (O'Keefe & Delia, 1978, 1979); as construct differentiation increases, so does the comprehensiveness of an individual's constructs.

Research on construct abstractness and comprehensiveness shows that perceivers who are relatively advanced in construct differentiation do not simply form more elaborate and organized impressions of others; they form qualitatively different kinds of impressions. As construct differentiation increases, so does the tendency to assess others along dimensions that are less tied to specific conditions of observation (for example, the particularities of behavior, the perceiver's interactional goals, or the social role that the target happens to occupy).

Because of this qualitative difference, Delia et al. (1975) hypothesized that perceivers who differed in construct differentiation would also differ in the degree to which their impressions and evaluations of a target are relatively context-independent or influenced by the dominant activity or goal involved in anticipated interaction with a target. They found that given a set of evaluatively inconsistent trait adjectives describing the work-related and social behavior of a fictitious target, perceivers low in construct differentiation formed evaluations consistent with the valence of the work-related attributes when told that the target would be a work partner. Their evaluations were consistent with the valence of the social attributes when told that the target would be a partner in informal social interaction. The evaluations of perceivers high in construct differentiation, by contrast, generally reflected an effort to integrate both kinds of information regardless of the kind of interaction they anticipated having with the target.

Taken together, previous research on the effects of perceiver construct differentiation on impression structure and content led to my expectation that, as students in my working groups became acquainted with each other and elaborated their impressions, group members who were high in construct differentiation (as compared to their less differentiated coworkers) would form impressions of their work partners that were more differentiated, more highly organized, and that reflected the use of more context-independent and fewer context-dependent dimensions of judgment.

Construct Differentiation as a Target Characteristic: Impact on Impressions

Developments in the interpersonal construct system (and especially increasing construct differentiation) have been found to have substantial influence on the way people behave in interpersonal settings, as well as on the manner in which they acquire and process information about others. Two lines of research have shown that increases in construct differentiation are associated with (a) increasing complexity in interpersonal choices and (b) increasing complexity in the design of interpersonal messages.

Construct differentiation and interpersonal choices. A recent line of work by O'Keefe and his coworkers (see O'Keefe, i 980) suggests that increases in construct differentiation are related to a decreased likelihood of behaving in a simple, attitude-consistent manner toward others. Using a framework derived from Fishbein's (Fishbein & Azjen, 1975) model of the attitude-behavior relationship, O'Keefe predicted that when subjects were asked to specify the likelihood of their undertaking each of a set of behaviors (some expressing positive affect toward another and some expressing negative affect), low-differentiation perceivers would make more attitude-consistent choices than high-differentiation perceivers. His work has shown that high-differentiation perceivers are in fact less dominated by simple evaluative consistency rules, and that their behavioral choices are substantially more variable and attitude-inconsistent than those of low-differentiation perceivers.

Construct differentiation and message design. Delia and his coworkers have produced a line of work that explores the relationship between developments in the interpersonal construct system and developing strategies for managing complex communication tasks through verbal message design. A complex communication task is one in which the message producer's primary objective in communicating is blocked by some obstacle in the situation (such as a target's unwillingness to comply with a request), or in which the message producer's aims in communicating are contradictory or incompatible (for example, a supervisor needs to correct the behavior of an erring subordinate but does not want to hurt the subordinate's feelings). In summarizing this line of work, O'Keefe and Delia (1982) concluded that whereas high-differentiation perceivers respond to complex communication tasks by producing messages that are more complex (containing features designed to address obstacles or subsidiary communicative objectives such as identity management), low-differentiation perceivers are more likely to produce simple messages that do little more than express and pursue the message producer's primary aim.

These two lines of work suggest that in general, the behavior of high-differentiation subjects should be less consistent and less transparent than the behavior of low-differentiation subjects, since high-differentiation subjects are less dominated by simple principles of behavioral organization (such as evaluative consistency rules or a focus on some simple communicative objective). The task of forming an integrated impression of a high-differentiation target should therefore place greater demands on a perceiver than the task of construing a low-differentiation target.

This led to my expectation that impressions formed of high-differentiation targets would exhibit higher levels of organization than impressions of low-differentiation targets. As described above, the level of organization of an impression reflects the degree to which the perceiver has recognized, represented, and reconciled variability or inconsistency in the behavior of a target. But research on the relationship of construct differentiation to interpersonal choices and message design suggests that with low-differentiation targets, there is less inconsistency or variability in the behavior to be construed. Therefore, as the target figure's construct differentiation increases, the target's behavior should exhibit increasing complexity. As a consequence, the target's construct differentiation should be related to the level of organization of impressions formed by perceivers about him or her.

Construct Differentiation and the Evolution of Impressions

As mentioned in the introduction to this chapter, the relevant research on the role of construct differentiation in influencing the information processing style of perceivers and the interpersonal behavior of targets has for the most part been conducted in non-naturalistic settings and in no case has explored the operation of these processes over the course of developing acquaintance. Thus there was little basis for predicting how the characteristics of perceiver and target would express themselves over time as in-teraction progressed.

Given the consistent and stable relationship between perceiver construct differentiation and impression characteristics across a wide range of different kinds of impressions (impressions of fic-titious and real targets, impressions formed from different kinds of provided information, impressions formed under different in-structional sets, and so on), one would expect that perceiver con-struct differentiation would have a continuing effect of stable magnitude on impressions over time. By contrast, the effects of target construct differentiation on level of organization should in-crease with time as the complexity of the target's behavior becomes manifest through a sequence of actions.

Although the primary concerns of this project were with the ef-fects of perceiver and target construct differentiation on impres-sions, there was also some basis in previous research to expect that independent of effects of construct differentiation, impres-sions would conform to an orderly pattern of development. Based on the observation that people have more differentiated impres-sions of those they like than those they dislike, Crockett (1965) suggested that impression differentiation is a function of amount of interaction as well as perceiver characteristics. This explana-tion implies that as interaction continues, impressions formed by all perceivers should become more differentiated.

Research on primacy-recency effects in impression formation (Klyver, Press, & Crockett, 1972; Mayo & Crockett, 1964) suggests that as evaluatively inconsistent information is presented sequen-tially, high-differentiation perceivers are more likely than low-differentiation perceivers to integrate early and late information and thereby form more organized impressions. This work suggests that as interaction continues, the impressions of high-differentiation perceivers (and perhaps, to a lesser extent, those of low-differentiation perceivers) should become more highly organized.

There is little or no foundation in previous research for any ex-pectation regarding general patterns of change in impression con-tent. Studies such as Duck (1973), which find that different kinds of constructs become important for judging friends as relation-ships develop, have only the most tangential relevance to the ques-tion of how impression content shifts during the development of a work relationship. Nonetheless, common sense would suggest that as people routinely interact within a work-oriented context, their impressions of each other should increasingly reflect the kinds of behavior they observe (which, after all, will be behavior appropriate to the work situation) and the dimensions of judgment that are relevant to the target's task functioning.

In general, one would expect that in a group work situation and over time, impressions would become increasingly differentiated and organized, and come to contain greater numbers of task-relevant beliefs about targets. The perceiver's construct differentiation should influence the kinds of impressions formed by consistently leading to relatively higher differentiation, relatively higher levels of organization, and proportionately fewer strictly task-relevant beliefs. The target's construct differentiation should influence the level of organization of impressions, since the level of organization of an impression depends on the recognition of complexity in a target's behavior, and the complexity of a target's behavior in turn depends on the target's construct differentiation.

IMPRESSION FORMATION IN WORKING GROUPS

This section describes and presents results of an initial investigation designed to explore the effects of perceiver and target construct differentiation on natural processes of impression formation. In this investigation, undergraduate students were assigned to work in groups on a graded project. At three points during their work on the project, students were asked for their impressions of each of their fellow group members. The structure and content of these impressions were analyzed, the effects of perceiver and target construct differentiation on impression structure and content were assessed, and patterns of change in impressions over time were identified.

Subjects

This project was conducted in two introductory small group communication classes at Wayne State University. In all, 44 students (20 in one class, 24 in the other) participated in the project. During the first week of class and before being assigned to project groups, students completed a two-peer version of Crockett's (1965) Role Category Questionnaire. Impressions written in this questionnaire were scored for construct differentiation using procedures outlined by Crockett et al. (1975). These impressions were scored by the experimenter, a trained coder who had achieved reliabilities of r = .95 to .99 in scoring RCQs with a variety of other coders in previous investigations.

Students' RCQ scores (within class) were rank-ordered and split at the median to produce two subject groups: a high-differentiation and a low-differentiation group. In assigning students to groups, the instructor randomly assigned two high-differentiation and two low-differentiation students to each group. Unfortunately, the instructor's assignment of students to groups was not preserved for all groups. Students were assigned to groups prior to the close of the initial enrollment period for the term. Subsequent to the asssignment of students to groups, there were some shifts in group membership as students added and dropped the class. This resulted in one group containing three low-differentiation perceivers and one high-differentiation perceiver and two groups containing three high-differentiation perceivers and one low-differentiation perceiver. Mean RCQ differentiation scores for each group thus ranged from 20.00 to 30.50, although means for nine of the groups fell between 21.25 and 25.50. In assigning students to groups, the instructor took care to ensure that students assigned to each group were unacquainted with each other prior to the start of the term.

Member Perception Questionnaire

Impressions of fellow group members were obtained from each subject at three points through repeated administration of the Member Perception Questionnaire (MPQ). This questionnaire was completed (a) on the class day immediately following the initial group meeting (during the second week of class), (b) after groups had worked together for two weeks on their projects (during the fourth week of class), and (c) following the presentation of reports to the class (during the sixth week of class).

The MPQ asked the student to identify and describe, in a short written paragraph, each of his or her fellow group members. Instructions for the MPQ paralleled those on the Role Category Questionnaire. In describing each group member, subjects were asked to describe each person so that someone who was unfamiliar with the target would understand what the target was like. Subjects were specifically asked not to provide physical descriptions of targets.

Analysis of Group Member Impressions

The differentiation and level of organization of each impression a subject provided over the course of the investigation was determined. In scoring impressions for differentiation, a distinction was drawn between beliefs relevant to the immediate social enterprise (working in a group on a particular kind of task) and non-context-relevant beliefs. For each impression, a subject thus received two differentiation subscores (task differentiation and non-task differentiation) which were summed to produce a total differentiation score. The analysis of each impression involved four steps: (1) segmentation of the impression into its separable belief elements (constructs); (2) classification of each construct into one of two categories, task- or non-task-relevant; (3) counting the number of elements in each category and summing these subscores to produce a total differentiation score; and (4) analysis of the overall structure of the impression to determine its level of organization.

Segmentation of impressions. Impressions were segmented using the same procedures employed in identifying separable constructs in the RCQ impressions.

Classification of constructs. Each construct employed in an impression was classified as either task-relevant or non-task-relevant using a set of simple coding rules. The focus of these rules was on the delineation of the task-relevant category. Any construct not assigned to the task-relevant category was automatically included in the non-task-relevant category. A conservative scoring rule was adopted: Only constructs that were clearly and unambiguously task-relevant were assigned to that category.

Constructs were classified as task-relevant when: (a) they described or referred to behaviors produced in pursuit of the group's assigned project (for example, "offered a lot of good ideas about our project," "always came to meetings on time," "had good eye contact during the presentation"); (b) described or referred to abilities of the target relevant to the group's task ("had leadership abilities," "intelligent," "knowledgeable," "well-organized"); (3) described or referred to dispositions of the target related to task performance ("the group's performance was important to him," "reasonable," "efficient," "hard-working"); or (4) described or referred to abilities and dispositions not naturally linked to task performance, but tied to the task context by a contextualizing phrase or qualifier ("mature in handling group responsibilities," "interested in the ideas of fellow group members," "open to ideas and suggestions"). Without these contextualizing phrases, "mature," "interested," and "open" were assigned to the non-task-relevant category.

Two coders independently scored 31 randomly selected impressions for the number of task-relevant constructs contained in each impression, with exact agreement for 81 percent of the impressions; the correlation for interrater reliability was $r = .91$.

Level of organization. Each impression a subject wrote was also scored for its overall level of organization. Level of organization scores reflect the degree to which inconsistency and variability in the behavior of a target is recognized, represented, and integrated into a consistent and coherent depiction of the target's character and motivations. Impressions were scored for level of organization using the system developed by Crockett et al. (1975) for assigning naturally formed impressions to one of fifteen levels of organization. These fifteen levels reflect minimal, average, or advanced use of one of five hierarchically ordered methods of information integration: (1) aggregation, in which no variability in the behavior of a target is represented; (2) implicit recognition of variability through the use of qualifiers, oppositional prepositions ("but," "however"), or contextual or motivational constraints modifying dispositional attributions; (3) explicit recognition of variability, without an account of the variability (for example, "He can be helpful, but he can also be selfish"); (4) resolution of variability, in which some explanatory mechanism is explicitly offered to account for variability in the target (either internal, such as a dynamic motivational state like insecurity used to account for variations in target friendliness, or external, such as specifying how variations in social environment lead to differences in manifest extroversion); and (5) extended resolution of variability, in which variations in target behavior are represented and explained as a function of both internal and external factors.

As in the case of differentiation, impressions were scored for level of organization by the experimenter, a trained coder who had achieved reliabilities of $r = .82$ to $.91$ in previous applications of this scoring system.

Construction of Dependent Measures

The analysis of group member impressions written by a subject served as the basis for constructing a number of indices of
impression structure and content.

Scores assigned to impressions could not be subjected directly to quantitative analysis due to problems created by missing data
and subject attrition. In all, 39 students completed the first MPQ
questionnaire, 36 the second, and 39 the third. Of the original 44
subjects initially included in the project, 28 produced complete
sets of data. The impressions they wrote in the MPQs reflected
any changes in group composition that occurred over the course
of the project. As a consequence, the number of impressions written about stable group members (members who had been
continuing participants in group activities from the inception of
the project to its conclusion) varied from group to group.

All analyses reported here were based on data produced by the
28 complete cases. In order to permit meaningful comparisons
across these subjects, only impressions of stable group members
(those who participated in the project from beginning to end) were
employed in computing the following indices of impression structure and content: (1) task differentiation—the mean number of
task-relevant constructs contained in a given perceiver's impressions at a given time; (2) non-task differentiation—the mean
number of non-task-relevant constructs contained in a given
perceiver's impressions at a given time; (3) total differentiation—
the sum of task and non-task differentiation (reflecting the mean
total differentiation in a perceiver's impressions at a given time);
and (4) level of organization—the mean level of organization of
a perceiver's impressions written at a given time. These four
measures were computed for each perceiver at each of the three
times the MPQ was completed (initial, second, and final).

Analysis and Results

A variety of analyses were performed in order to assess the effects of perceiver and target construct differentiation on impressions. In assessing the effects of target construct differentiation
on impressions, the mean RCQ score of targets about whom a
perceiver wrote impressions was used as a predictor variable. For
some analyses used to assess effects of perceiver RCQ on impres-

sions, perceivers were split into two groups (high and low differentiation) by rank-ordering their RCQ scores and splitting the distribution at the median value. Discussion of the analyses and results is organized in two sections: (1) effects on differentiation (total, task, and non-task); and (2) effects on level of organization.

Effects on differentiation: total, task, and non-task. Contrary to expectation, total differentiation was not substantially related to perceiver RCQ scores and did not increase linearly with time. A 2 x 3 repeated-measures analysis of variance on total differentiation scores in which the between-subjects factor was perceiver construct differentiation (two levels: high and low) and the within-subjects factor was time (three levels: initial, second, and final) showed only one significant effect: a main effect for time, $F (2, 52) = 4.15$, $p< .05$. The highest total differentiation scores were produced in the initial impressions ($M = 6.69$), with a decline in differentiation in the second impressions ($M = 5.73$) and a slight increase from the second to the final impressions ($M = 6.09$).

The zero-order correlations between perceiver RCQ scores and total differentiation at each time, moreover, showed that while perciever RCQ differentiation had a marginally significant effect on the total differentiation of initial impressions ($r = .327$, $p< .09$), it had no significant effect thereafter. The correlations between perceiver RCQ differentiation and total differentiation were not even marginally significant for the second and the final group member impressions.

The reason that perceiver construct differentiation was essentially unrelated to total differentiation in group member impressions became clear when effects on task and non-task differentiation were analyzed separately. Perceiver construct differentiation was essentially unrelated to task differentiation at any time, but the correlation between percerver construct differentiation and non-task differentiation was significant for the first impressions ($r = .469$, $p< .05$) and marginally significant for the second impressions ($r = .3544$, $p< .07$). However, this correlation was near zero and not significant for the final impressions ($r = .10$, $p< .61$).

The failure to find a significant correlation between perceiver construct differentiation and non-task differentiation in the final set of impressions reflects the fact that, at the time of the final impression, perceivers tended to represent their fellow group members along task-relevant dimensions or along non-task relevant dimensions, but not both. In the first and second set of group member impressions, the correlation of task and non-task differentiation was essentially zero (respectively, $r = -.17$ and $r = .04$), but for the final set of impressions, task and non-task differentiation was significantly and negatively correlated ($R = -.482$, $p < .01$).

The effects of time on total differentiation were also further clarified by contrasting the effects of time on task and non-task differentiation. A $2 \times 3 \times 2$ repeated-measures analysis of variance in which the between-subjects factor was perceiver construct differentiation (two levels, high and low) and the two within-subjects factors were time (three levels: initial, second, and final) and differentiation type (two levels, task and non-task)revealed a number of significant effects. There was a significant time-by-differentiation type interaction, $F(2, 52) = 5.66$, $p < .01$, which contained a significant main effect for time, $F(2, 52) = 4.15$, $p < .05$. There was also a nearly significant perceiver construct differentiation by differentiation type interaction, $F(1, 26) = 3.59$, $p < .07$. A significant main effect for differentiation type, $F(1, 26) = 8.69$, $p < .01$, was contained within each of these interactions.

The main effect for time was the same effect reported previously in the discussion of total differentiation: Differentiation declined from the first impressions to the second and then increased slightly from the second to the third. The main effect of differentiation type reflected the fact that disregarding time, impressions contained fewer task-relevant than non-task-relevant constructs. But the interaction between time and differentiation type was due to the fact that whereas task differentiation was about equal in the first and second impression sets and increased in the third impression set, non-task differentiation decreased substantially from the first to the second impression set and then decreased further from the second to the third set (see Table 10.1). The elaboration of context-relevant beliefs, combined with the decline in numbers of non-context-relevant beliefs, suggests that the initial representation of targets in terms of general, task-independent dimensions of judgment was increasingly supplanted by task-specific judgments as work in the groups progressed.

However, as predicted, high- and low-differentiation perceivers differed in the degree to which they elaborated task-independent representations of targets. High and low-differentiation perceivers used very nearly equal numbers of task constructs in describing fellow group members; the interaction between perceiver construct differentiation and time is attributable to the greater number of non-task constructs employed by high-differentiation perceivers (see Table 10.1).

TABLE 10.1   Mean Numbers of Task-Relevant and Non-Task-Relevant
Constructs as a Function of Perceiver Construct
Differentiation and Time

	Time 1	Time 2	Time 3	Mean Across Time
High-differentiation perceivers				
task-relevant constructs	2.34	2.31	2.75	2.47
non-task-relevant constructs	4.89	3.65	3.48	4.01
Low-differentiation perceivers				
task-relevant constructs	2.44	2.48	3.32	2.74
non-task-relevant constructs	3.42	2.89	2.55	2.95

It was not expected that target construct differentation would have significant effects on task, non-task, or total differentiation, and it had none. The zero-order correlations of target construct differentiation with these three variables were nearly zero; regression analyses performed to determine whether target and perceiver construct differentiation interacted to influence any of these three variables similarly showed no significant effects.

Effects on level of organization. Results for level of organization showed that while the effect of target construct differentiation on level of organization increased and became significant as interaction progressed, there were no other significant effects on level of organization. In the first impression set, the zero-order correlation of target construct differentiation and level of organization was essentially zero ($r = .0358$); however, in the second impression set the correlation was $r = .2366$, $p < .12$, and in the third impression set the correlation was substantial ($r = .46$, $p < .01$). Regression analysis performed to determine whether perceiver and target construct differentiation interacted in influencing level of organization showed no significant interaction. The zero-order correlations between perceiver construct differentiation and level of organization in each of the impression sets were nonsignificant, and there were no significant effects in a 2 x 3 (perceiver construct differentiation by time) repeated-measures analysis of variance on level of organization.

DISCUSSION

Before going on to discuss the implications of these findings, I want to offer one caveat: Conclusions drawn purely on the basis of these findings ought to be drawn very carefully, since the sample was quite small, and some important sources of variance (such as group character and culture) could not be taken into account due to imbalances in the design created by subject attrition. Because the project was exploratory and descriptive in focus, the design of the study did not reflect an effort to carefully isolate perceiver and target processes as causal influences on impressions. In this context, it is important to report that two of the major findings of this project were subsequently replicated under less equivocal conditions.

In a recent investigation, O'Keefe and Shepherd (1983) asked subjects who were previously unacquainted to engage in dyadic persuasive interactions and to provide impressions of their partners after interacting. As in the present investigation, the level of organization of postinteraction impressions was significantly and positively correlated with the target's score on the Role Category Questionnaire. O'Keefe and Shepherd also classified the constructs contained in impressions of interactional partners as to their situational relevance using a refined version of the construct coding system outlined in this chapter. They found that perceiver construct differentiation (as assessed by the RCQ) was not related to task-relevant differentiation or to the number of constructs used in describing the affiliative displays and politeness of the target ("relational differentiation"). However, it was strongly and positively related to the number of constructs used to describe context-independent features of the background and personality of the target (a category of constructs O'Keefe and Shepherd labeled "character analytic").

A second investigation by Delia and Murphy (1983) provided an additional replication of the finding that the level of organization of the impression a perceiver forms is related to the target's level of construct differentiation. In their study, the target's college roommate was asked to provide an impression. They found that the level of organization of the roommate's impression was significantly and positively correlated with the target's RCQ score.

Despite the limitations of the design and analyses employed in this initial investigation, the results obtained appear to be reliable and generalizable beyond the context of classroom group interaction. In discussing the findings of the present investigation, I first consider their implications for conceptions of the role of perceiver construct differentiation in impression formation and then for conceptions of impression accommodation.

Perceiver Construct Differentiation and Impression Formation

Perhaps the most consequential result of this project was the failure to find any significant correlation between perceiver's RCQ construct differentiation and the total differentiation of group member impressions. Instead, RCQ differentiation was significantly related only to the number of non-task-relevant constructs perceivers employed. Subjects who differed in RCQ differentiation did not differ in the numbers of task-relevant constructs they employed. In O'Keefe and Shepherd's (1983) subsequent study, impression differentiation was similarly segmented into components: task-relevant (involving assessments of the target's skill as a persuader and arguer, since the assigned task was to discuss a policy issue), relationship-relevant (involving assessments of the target's consideration and affiliative behavior), and character analytic (involving global evaluations of the target and assessments of the target's social background, moral character, and non-situationally relevant dispositions such as generosity). In O'Keefe and Shepherd's study, as in the present investigation, the various components of differentiation were not intercorrelated. Moreover, RCQ differentiation was significantly related only to character-analytic differentiation, the component that had the least obvious relation to the immediate goals of perceiver and target.

These findings immediately suggest two hypotheses: (1) that beliefs reported in free-response impressions arise from the operation of more than one cognitive mechanism or process (the independence of differentiation components provides support for this hypothesis); and (2) that differences in impression-formation style associated with RCQ differentiation are qualitative rather than quantitative in origin (the association of RCQ scores with one specific component of impression differentiation provides support for this claim). Neither of these hypotheses is consistent with a view of RCQ scores as reflecting, in some direct fashion, the number of constructs in a perceiver's interpersonal construct system. (In Crockett's [1965] initial development of this instrument and scoring procedures for it, it was represented as just such an assessment.)

In a recent survey of research on the relationship of construct differentiation to interpersonal behavior, O'Keefe and Delia (1982) explicitly argue that this original interpretation of RCQ scores is probably incorrect. They offered two plausible alternative interpretations of RCQ scores: First, that scores on the RCQ might reflect the number of abstract (dispositional) constructs a person has available for construing targets (an interpretation suggested by recognizing that the instructions for the RCQ create a set to describe abstractly; the scoring procedures for the RCQ involve exclusion of some kinds of more concrete constructs; and the correlations between RCQ differentiation and construct abstractness are quite high). Second, they have suggested that scores on the RCQ might reflect the number of beliefs a person can quickly retrieve about a target (and thus the organization of beliefs), rather than the number of beliefs a person actually has or the number of constructs used in forming beliefs.

In light of the present findings, neither of these suggestions seems very satisfactory. Both of O'Keefe and Delia's alternative suggestions were designed to account for purely quantitative differences in impression differentiation, and both are premised on the assumption that impressions are produced through the operation of a coherently organized and uniformly developing social cognitive system. By contrast, the results of this project and of O'Keefe and Shepherd's (1983) study suggest that beliefs relevant to different aspects of a target's performance in a given social situation are produced through the application of different sets of constructs (some used to assess performance on the particular task being performed in the situation, some used to assess interpersonal demeanor, and some with no obvious relevance to any specific feature of the situation but only to a more general understanding of character and motivation). Differentiation in only one of these sets of constructs (situation-independent or character-analytic) is associated with performance on the RCQ.

Some time ago, Hastorf, Richardson, and Dornbusch (1958) offered a number of conclusions drawn from systematic examination of free-response impressions, one of which was that in describing a range of targets, "a person has a core of generally consistent categories used in describing all people and a set of more particular categories which depend more on situational factors [for their use]" (p. 61). This observation is quite consistent with present findings and suggests that the differential elaboration of such a set of core constructs is responsible for differences in RCQ scores.

The interesting question raised by this interpretation of RCQ scores and the character-analytic or non-task-relevant component of impressions as reflecting the operation of some set of core constructs is why (if they are truly "core" constructs) the number of non-task-relevant constructs in group member impressions declined over time. One would expect that it is precisely core constructs that should always be employed in judgment and thus form the stable core of an impression.

On the assumption that such a set of core constructs exists, we know that its application to judgment is reflected in impressions of important and well-known others (such as those described in the RCQ) and in initial impressions of a wide range of targets (either fictitious or real). Inspection of the kinds of constructs that embody this hypothetical core (evaluations of the target's general quality as a person, assessments of moral character, and dispositions that are relevant to neither interpersonal demeanor nor practical tasks) suggests that this core of constructs is primarily functional for making decisions about relationships and, in particular, whether to pursue or avoid a close relationship with a target. These are exactly the kinds of asssessments one would expect to find in descriptions of friends, enemies, and close acquaintances (exactly the sort of people described in the RCQ) and in first impressions (where the trajectory a relationship will take is still an open question). These are also the kinds of assessments that should become increasingly less relevant in a situation like the one in this investigation. Relations between group members were highly constrained by the social structure in which they found themselves. They had little freedom to expand or extend their relations with each other to interaction not directly focused on the immediate task. Over the course of the project, subjects came increasingly to be simply task partners to each other. In short, the situation channeled relationships into a relatively narrow trajectory. Impressions written over the course of development of these relationships came increasingly to reflect the task-focused and interpersonally superficial quality of these relationships. The number of task-relevant constructs increased, and the number of non-task-relevant constructs (associated with RCQ scores) declined.

Several conclusions can be drawn from this analysis: (1) interpersonal construct systems should not be viewed as coherently organized and uniformly developing cognitive systems, but rather as a collection of interpretive structures, each of which is designed to perform some particular assessment function; (2) some assessment functions are specific to tasks (such as assessing the performance of a fellow group member or of a persuader), some are specific to important dimensions of social situations (such as assessing the degree to which a target behaves in a friendly and considerate manner), and some are specific to relational decision making (such as assessing the quality and character of a target along core value dimensions); (3) a perceiver's employment of a given interpretive structure (set of constructs) will depend on the relevance of the assessment function it serves to the conditions of observation; (4) increasing or decreasing elaboration of beliefs in relation to some specific assessment function will depend on the continuing relevance of that assessment function to the interaction between perceiver and target; (5) RCQ scores supply a quantitative index of a qualitative difference among perceivers, since they result form the differential elaboration of a particular interpretive structure (a set of core constructs that appear to be primarily useful for relational decision making) and not of the construct system as a whole.

Level of Organization and Impression Accommodation

A second important finding of this project was that the level of organization of group member impressions was significantly related to the mean RCQ score of the targets a given perceiver described. This same relationship was subsequently found in O'Keefe and Shepherd's (1983) study and in the study of roommate impressions by Delia and Murphy (1983). There was good reason to expect this effect, since RCQ scores have been found to be good predictors of a person's behavioral variability and complexity, and since level of organization scores reflect the perceiver's success at recognizing and explaining variability and complexity. The results of this study and subsequent replications thus suggest that as a person's RCQ score increases, the complexity of his or her behavior and qualities as a stimulus also increases. This is then reflected in the organization of impressions written about them.

The findings of this project stand in rather sharp contrast to most of the research on information integration in impression formation. The bulk of work on information integration has not been concerned with the ways people develop substantive representations of others, but only with the determinants of overall attitudes or evaluations (for example, Anderson, 1968). Research that has focused on the themes and structures used in organizing beliefs about others, like the work of Crockett and his associates, has treated the presence of such structures as simply expressive of the perceiver's underlying cognitive structures. While there is evidence that, at least under some circumstances, the use of such organizing themes and structures is governed by perceivers' skills at social perception, the use of higher-level strategies for information integration is motivated by the kind of information perceivers receive. In short, the level of organization of an impression reflects the perceiver's effort to accommodate to the real characteristics of the target in developing a representation.

This finding stands in contrast not only to the view that impression organization is simply a function of a perceiver's social cognitive skill, but also to the rather substantial body of work suggesting that perceivers are generally unresponsive to target characteristics in forming impressions. Much of this work has focused on the literal accuracy of perceivers' assessments of targets, or on the degree to which perceivers use specific dimensions of judgment in such a way as to discriminate among targets (see Schneider, 1973). Overwhelmingly, such studies have found that perceivers' ratings of different targets are dominated by target-independent relations among rating dimensions.

The findings of the present investigation suggest that the failure to find discriminating and responsive ratings of targets should not be taken as conclusive evidence that perceivers do not represent and respond to differences among targets. In fact, the results of this study suggest that the content of impressions (the kinds of beliefs an impression is likely to contain) reflect the perceiver's accommodation to the conditions of observation (the kind of social situation, the goals one has in that situation, and the kind of behavior one is observing). The fact that independent components of impressions can be identified by their relevance to different features of a social situation further suggests that what we call implicit theories of personality are in fact a collection of interpretive structures, each defined by its functional relevance for providing information in some particular kind of situation or in reference to some personal goal. This conjecture is supported by research on the organization of implicit theories of personality. For example, Rosenberg and Sedlak (1972) reported that two primary dimensions underlie the English trait vocabulary. Trait words useful for describing task behavior (such as "organized" or "lazy") formed opposing clusters (one positively valenced, one negatively valenced), as did trait words useful for describing interpersonal demeanor (such as "friendly" or "rude"). While Rosenberg and Sedlak interpret these clusters as reflecting fundamental dimensions of personality ("task intelligence" and "social intelligence"), it is equally possible to see these dimensions as defined by the functional relevance of the kind of assessment they provide for different kinds and features of situations.

The results of the present investigation certainly suggest that rather than having unitary "theories of personality," perceivers have a set of functionally organized interpretive structures, each of which is relevant to some judgment task, and the use of which is determined by the relevance of that judgment task to circumstances of observation. If this conjecture is correct, it is unreasonable to expect the structure or use of implicit theories of personality to be responsive to differences in targets, since the interpretive system of the perceiver is organized to be responsive to differences in situations.

If patterns in trait attribution (and therefore the content of impressions) primarily reflect the relevance of different kinds of judgment to the social situation, then it is unlikely that one will find accommodation to target characteristics by looking at the accuracy of specific judgments or at patterns of correlations among rating dimensions. Results of the present investigation suggest that accommodation to target characteristics is not to be found here, but rather can be found in the way in which specific assessments are combined in forming integrated representations. For example, in this study perceivers actually represented more complicated people as being more complicated. It is quite possible that a more refined analysis, one that examines the kinds of themes used in organizing representations of different kinds of people (rather than crudely coding for the presence or absence of such themes), would find further evidence of sensitive discrimination among targets and responsive adjustment of impressions to their characteristics.

CONCLUSION

While the results of this investigation have a number of specific implications for models of the interpersonal construct system and its operation in the process of impression formation, they also point to the dangers inherent in taking the man-as-scientist metaphor too seriously.

It is no exaggeration to say that thought about social cognition has been dominated for some time by the implicit scientist metaphor. Perceivers' cognitive structures are described as "theories." Everyday procedures for acquiring information are described as "hypothesis testing" and "experimentation." Perceivers' goals are assumed to be the goals of science: developing an accurate representation of reality along with the ability to predict, explain, and control events.

The fruitfulness of this metaphor arises precisely from its ability to provide a vocabulary for discussing knowledge, since scientific enterprises represent the most extensively described and carefully analyzed examples of knowledge systems and their rational properties. One reason that scientific enterprises offer such convenient and compelling cases for the analysis of problems related to epistemology is that the dominating purpose of science is to produce knowledge. This is a distinctive property of science (and other scholarly enterprises), as compared to other kinds of human endeavor. To insist (as so many theorists do) that all human enterprises are alike in producing knowledge is to mistake the important difference in priorities that makes scientific enterprises what they are.

It should be obvious that while human beings, whether in their capacity as scientists or in some other capacity, are constantly producing knowledge, scientists' activities are directed by the primary goals of producing knowledge, whereas most everyday practical knowledge is produced as a byproduct of human existence. Yet the implicit scientist metaphor implies that knowledge of people is a product, rather than a byproduct, of human life. That is, the acquisition and organization of knowledge about other people and social life is treated as if it were an end in itself, and the process of acquiring practical social knowledge is treated as an enterprise akin to pure (not applied) science: isolated from requirements of relevance, insulated from the exigencies and exertions of life, observational rather than interactional.

What I want to suggest is that the machinery of social cognition is organized around personal and social purposes rather than epistemological aims. People do not acquire knowledge for its own sake. They acquire it because they must in order to participate in social processes. People learn to organize and categorize behavior in social situations because performance in social situations requires them to know what forms behavior can take in those situations. People learn to evaluate the performance of others in relation to the normatively prescribed purposes of a situation because such assessments are important to achieving their practical goals. Learning to produce and interpret social behavior is a great deal more like learning a language than it is like developing a theory of the structure of matter.

Similarly, developing an impression of a target is a lot more like assigning an interpretation to a turn at talk than it is like discovering the properties of some physical substance. The guiding problem for the perceiver is to determine what behavior means and how it fits into a (presumably) coherent theme. And just as, in interpreting discourse, the meaning of an utterance will be assigned in relation to the goals that are taken as operative in a given situation, so in forming impressions the meaning of a target's behavior will be assigned in relation to interactional goals. Moreover, one's impression of a target will reflect the fact that, as with turns at talk, one interprets whatever one encounters because it is there and for the sake of knowing what has been done; behavior is not taken as an observational instance against which to test a theory of people in general or of the target in particular. Our knowledge of others is not developed for its own sake, in an effort to develop accurate representations, but as a byproduct of interaction. We deal with others for all kinds of reasons in the course of everyday life, and in the course of our dealings we pick up information about them along the way.

In short, in everyday life the dominating purposes of social cognitive structures are the assignment of interpretations that make behavior coherent and the functional assessment of behavior in relation to situated purposes—not the development of a sound theory of human nature or an accurate and sensitive representation of target personality. Recognizing that human beings are dominated by considerations of relevance, are primarily responding in terms of the exigencies of situations, and gain knowledge as a byproduct of interaction rather than a product of observation makes a great difference to what one expects in examining the organization and products of social cognitive processes. Rather than expecting to find organized "theories of personality," one would expect to find specialized and functionally independent interpretive structures, useful for doing some practical job of assessment. Rather than expecting to find accurate reflections of targets' characteristics in perceivers' beliefs, one would expect to find rough-and-ready assessments of targets and their behavior in relation to personal goals and values. Rather than expecting impressions to look like the output of a machine made to sensitively register characteristics of targets along a number of dimensions and form an accurate overall assessment, one should expect to find that impressions are cobbled together from the output of many different machines with diverse design principles, all activated by their relevance to the immediate job at hand.

All of this suggests that the key problem in the study of social cognition is the problem of *relevance*. I have argued that cognitive structures are organized in terms of their relevance for practical tasks and activated by their relevance to situated purposes. The chief defect in the implicit scientist metaphor is that it implies that such practical goals and relations are subordinate in importance to epistemological aims; surely the rationality and organization of everyday knowledge can be characterized in a more accurate and sympathetic way.

REFERENCES

Anderson, N. (1968). A simple model for information integration. In R. P. Abelson et al. (Eds.), *Theories of cognitive consistency: A sourcebook.* Chicago: Rand-McNally.

Applegate, J. L., Kline, S. L., & Delia, J. G. (1980). *Alternative measures of cognitive complexity as predictors of communication performance.* Unpublished manuscript, University of Kentucky.

Asch, S. E. (1946). Forming impressions of personality. *Journal of Abnormal and Social Psychology, 41,* 258-290.

Biskin, D. S., & Crano, W. (1977). Structural organization of impressions derived from inconsistent information: A developmental study. *Genetic Psychology, 95,* 331-348.

Bruner, J. S., & Taguiri, R. (1954). The perception of people. In G. Lindzey (Ed.), *Handbook of social psychology: Vol. 2.* Reading, MA: Addison-Wesley.

Burke, J. A. (1979). *The relationship of interpersonal cognitive development to the adaptation of persuasive strategies in adults.* Paper presented at the Central States Speech Association Convention, St. Louis.

Crockett, W. H. (1965). Cognitive complexity and impression formation. In B. A. Maher (Ed.), *Progress in experimental personality research: Vol. 2.* New York: Academic Press.

Crockett, W. H., Gonyea, A. H., & Delia, J. G. (1970). Cognitive complexity and the formation of impressions from abstract qualities or from concrete behaviors. *Proceedings of the 78th Annual Convention of the American Psychological Association. 5,* 375-376.

Crockett, W. H., Mahood, S. M., & Press, A. N. (1975). Impressions of a speaker as a function of set to understand or to evaluate, of cognitive complexity, and of prior attitudes. *Journal of Personality, 43,* 168-178.

Crockett, W. H., Press, A. N., Delia, J. G., & Kenny, C. T. (1975). *Structural analysis of the organization of written impressions.* Unpublished manuscript, University of Kansas.

Delia, J. G. (1972). Dialects and the effects of stereotypes on interpersonal attraction and cognitive processes in impression formation. *Quarterly Journal of Speech, 41,* 119-126.

Delia, J. G. (1976). A constructivist analysis of the concept of credibility. *Quarterly Journal of Speech, 62,* 361-375.

Delia, J. G., Burleson, B. R., & Kline, S. L. (in press). The organization of naturally formed impressions in childhood and adolescence. *Journal of Genetic Psychology.*

Delia, J. G., Clark, R. A., and Switzer, D. E. (1974). Cognitive complexity and impression formation in informal social interaction. *Speech Monographs, 41,* 299-308.

Delia, J. G., Crockett, W. H., Press, A. N., & O'Keefe, D. J. (1975). The dependency of interpersonal evaluations on context-relevant beliefs about the other. *Speech Monographs, 42,* 10-19.

Delia, J. G., Kline, S. L., Burleson, B. R., Clark, R. A., Applegate, J. L., & Burke, J. A. (1980). *Social-cognitive and communicative skills of mothers and their children.* Unpublished manuscript, University of Illinois at Urbana-Champaign.

Delia, J. G., & Murphy, M. A. (1983). *Roommates' construct differentiation, impressions, and person-centered communication: An analysis of perceiver and target effects.* Paper presented at the annual meeting of the Speech Communication Association, Washington, DC.

Duck, S. A. (1973). *Personal relationships and personal constructs.* New York: John Wiley.

Fishbein, M., & Azjen, I. (1975). *Belief, attitude, intention, and behavior.* Reading, MA: Addison-Wesley.

Gollin, E. S. (1954/1955). Forming impressions of personality. *Journal of Personality, 23,* 65-76.

Hastorf, A. H., Richardson, S. A., & Dornbusch, S. M. (1958). The problem of relevance in the study of person perception. In R. Taguiri & L. Petrullo (Eds.), *Person perception and interpersonal behavior.* Stanford, CA: Stanford University Press.

Kelly, G. A. (1955). *A theory of personality.* New Yorks: W. W. Norton.

Klyver, N., Press, A. N., & Crockett, W. H. (1972). *Cognitive complexity and the sequential integration of inconsistent information.* Paper presented at the annual meeting of the Eastern Psychological Association.

Mayo, C. W., & Crockett, W. H. (1964). Cognitive complexity and primacy-recency effects in impression formation. *Journal of Abnormal and Social Psychology, 68,* 335-338.

Meltzer, B., Crockett, W. H., & Rosenkrantz, P. S. (1966). Cognitive complexity, value congruity, and the integration of potentially incompatible information in impressions of others. *Journal of Personality and Social Psychology, 4,* 338-343.

Nidorf, L. J., & Crockett, W. H. (1965). Cognitive complexity and the integration of conflicting information in written impresssions. *Journal of Social Psychology, 79,* 165-169.

O'Keefe, B. J., & Delia, J. G. (1978). Construct comprehensiveness and cognitive complexity. *Perceptual and Motor Skills, 46,* 548-550.

O'Keefe, B. J., & Delia, J. G. (1979). Construct comprehensiveness and cognitive complexity as predictors of the number and strategic adaptation of arguments and appeals in a persuasive message. *Communication Monographs, 46,* 321-340.

O'Keefe, B. J., & Delia, J. G. (1982). Impression formation and message production. In M. Roloff & C. Berger (Eds.), *Social cognition and communication.* Beverly Hills, CA: Sage.

O'Keefe, B. J., & Delia, J. G. (in press). Psychological and interactional dimensions of communicative development. In H. Giles, R. St. Clair, & M. Hewstond (Eds.), *Advances in language, communication, and social psychology.* Hillsdale, NJ: Lawrence Erlbaum.

O'Keefe, B. J., Delia, J. G., & O'Keefe, D. J. (1977). Construct individuality, cognitive complexity, and the formation and remembering of interpersonal impressions. *Social Behavior and Personality, 5,* 229-240.

O'Keefe, B. J., & Shepherd, G. J. (1983). *Defining the communication situation: Implications for perceptions and actions.* Paper presented at the annual meeting of the Speech Communication Association, Washington, DC.

O'Keefe, D. J. (1980). The relationship of attitudes and behavior: A constructivist analysis. In D. P. Cushman & R. D. McPhee (Eds.), *Message-attitude-behavior relationship.* New York: Academic Press.

O'Keefe, D. J., & Sypher, H. E. (1981). Cognitive complexity measures and the relationship of cognitive complexity to communication: A critical review. *Human Communication Research, 8,* 72-92.

Press, A. N., Crockett, W. H., & Delia, J. G. (1975). Effects of cognitive complexity and of perceiver's set upon the organization of impressions. *Journal of Personality and Social Psychology, 32,* 865-872.

Rosenberg W., & Sedlak, J. (1972). Structural representations of implicit personality theory. In L. Berkowitz (Ed.), *Advances in experimental social psychology: Vol. 6.* New York: Academic Press.

Rosenkrantz, P. S., & Crockett, W. H. (1965). Some factors influencing the assimilation of disparate information in impression formation. *Journal of Personality and Social Psychology, 2,* 397-402.

Scarlett, H. H., Press, A. N., & Crockett, W. H. (1971). Children's descriptions of peers: A Wernerian developmental analysis. *Child Development, 44,* 439-453.

Schneider, D. J. (1973). Implicit personality theory: A review. *Psychological Bulletin, 79,* 294-309.

Werner, H. (1957). The concept of development from a comparative and organismic point of view. In D. B. Harris (Ed.), *The concept of development.* Minneapolis: University of Minnesota Press.

11

Relationship Growth and Decline

Steve Duck, Dorothy E. Miell, and David K. Miell

Social behavior is all too often described, discussed, and explained as if it were a dry abstraction. Most attention is usually paid to the cognitive processes that "drive" it; for instance, when studies of development in personal relationships talk of similarity of attitudes, complementarity of personality, self-disclosure, and information exchange (see Morton & Douglas, 1981). In our view, such abstractions take too little account of the ways in which such cognitive features actually create their effects in social behavior. For example, similarity between two people can be an effective social influence on their relationship only if they can communicate that similarity to one another. Because laboratory experiments on such similarity always do this communicating for the subjects as an inherent part of the experimental technique, we learn nothing about whether and how subjects do it for themselves. We learn nothing about the skills necessary to reveal, discover, and inter-

Authors' Note: *We gratefully acknowledge the support of grant HR5382 from the Social Science Research Council during preparation of this chapter.*

pret the presence of similarity, and we necessarily have to assume optimal skills, optimal communication, and optimum equality of ability in interpretation by all subjects, equally.

The present chapter will consider research on the development of personal relationships from another perspective. We will attempt to balance understanding of the cognitive components (such as information about partner, representation of the relationship, empathic understanding of the partner, and intimacy of self-disclosure) with the social and communicative components (for example, the ways in which information is transmitted socially, the behavioral representation and communication of understanding, and the techniques of self-disclosing). Consideration will be given to the strategic uses to which self-disclosure is put in developing and maintaining relationships. Its use is not causal, nor motivated merely by desires for reciprocity, except when the relationship is known to be distant or unlikely to develop. It is used by people who wish to develop relationships, and it is used systematically. Second, we shall discuss the equally strategic use of other cognitive components of relationships, such as similarity and empathy. Study of each of these takes a different focus once it is viewed as part of a behavioral recipe for relationship creation rather than as a disembodied abstraction that creates relationships almost without action by the partners themselves.

RELATIONSHIP DEVELOPMENT: PROPERTIES OR PROCESSES?

The bulk of work on interpersonal attraction has been done by social psychologists and has concerned itself for many years with the point of initial attraction to a stranger. Many subjectively attractive characteristics, features, and tendencies were identified in such work, ranging from physical attractiveness (Perrin, 1921; Berscheid & Walster, 1974), to attitudinal geography (Byrne, 1971), to personality contours similar to those of the subject (Duck, 1977). The unexpressed implication of such work was that the identified experimental variable was a key to the attraction of one person to another in real life, and indeed there were studies devoted to expressing this point explicitly (for example, Byrne, Ervin, & Lamberth, 1970). The present section looks at some of the conceptual deficiencies of such work.

It was not always made clear in such work that single characteristics of a partner would be an absurd basis on which to form a relationship. Much more sensible is the suggestion that different features of the partners take over the motive force of the relationship once one sees the significance of time itself in the development of relationships. The first light of day dawned when a filter theory of courtship growth was postulated (Kerckhoff & Davis, 1962). In that theory, at least two features of the partners were thought to influence relationship growth: need complementarity and value consensus. Kerckoff and Davis proposed that, given a similarity of social characteristics, value consensus and need complementarity would foster progress towards permanence in courtship.

A filter theory of friendship, as distinct from courtship, was postulated by Duck (1973, 1977), arguing that friends proceed to develop their relationship by reference to a whole series of such filters. These begin with social or physical characteristics and end with similarity of those parts of personality that contain or concern the understandings of other people and how they "tick." The sequence is one which takes the acquainting partners to greater and subtler depths of knowledge about one another and, in so doing, provides them with a ready indication of the similarity, overlap, or contiguities between their two personalities at increasingly subtle levels. The process is thus conceived to provide a means of assessing one's own personal views of the world on a variety of possible dimensions and in a variety of categories. Essentially, then, it serves to offer validation for one's personality; the partner will remain attractive for as long as such validation is offered in a proportion and at a depth not provided by alternative companions.

In perceiving the relevance of series of cues in the development of relationships, such a view presents the researcher with a variety of interesting possibilities. On the one hand (after Hinde 1979, 1981), one could suppose that certain of the cues emerge in a relationship. Hinde uses this term in a special sense that has been misunderstood (see Levinger, 1980) to mean that certain aspects of a person take time to emerge. That is not what Hinde means. His claim is that certain properties that may be significant to relationship success are not the properties of either individual, but emerge instead from their coalescence. Thus, similarity is not a property of either partner alone, but something that emerges between them. Equally, intimacy does not "belong" to either partner, but emerges from their interaction. Clearly, then, research should focus on the ways in which emergent properties do emerge, the ways in which the emergence is construed by the partners and outsiders, and the means by which they exert their relational influence. In our view, it is totally inadequate and extremely simplistic to argue that the influence is due to the simple effects of time passing.

Duck and Sants (1983) have indicated ways in which the effects of time are too often overlooked in both the conceptualization of and research on relationships, arguing that these effects are not simply consequences of time passing. In an extended relationship, the relational partners do certain things that they had not gotten around to doing before and thus fulfill a more extensive range of goals and needs. They can assess and evaluate one another's behavioral style and cognitive apparatus. They can explore alternative forms of relationship, create new roles, establish changed patterns of activity, and so on. It is important, therefore, to include the effects of people having time in relationships. Such an approach indicates the missing components to which time provides the key, but to do this we must consider the pros and cons of filtering approaches.

While such approaches are an undoubted advantage over the early "static" models that suggest that relationships are simply triggered by the presence of a certain characteristic, they are clearly deficient explanations of the realities of relationship growth. People do not merely have to be similar, they have to know that they are; people do not automatically learn that they satisfy the relevant filtering criteria for intimacy growth, they have to actively find out that they do; people do not merely observe relevancies to their relationship's growth, they must develop a strategy that guides them; and, of course, they then need the skills to carry out such a stratgegy.

Once this view is adopted, it becomes clear that knowledge about one's partner is not enough (Berger & Bradac, 1982). The central feature of relationship growth is communication of and about that knowledge. A key part of such communication is that it should concern relevant parts of the total repertoire available, that it should focus on the parts of a person that are appropriate to the relationship depth that has been reached, and that it should be guided by and sensitive to the partner's response to one's own communication. Finally, however, it must be done in the skillful ways that social norms for relationships require, both in terms of pacing intimacy growth and in terms of depth probing.

Such an approach to the development of relationships, then, begins to emphasize the behavior of acquainting partners—not as distinct from the cognitions and cognitive geography that they are communicating about, but as an addition and complement to it. We argue here that individuals may be rejected because they are not similar to another or because they communicate their similarity inexpertly, inappropriately, or out of time. To take cognitive similarity out of its behavioral, strategic context is to provide only the ingredients and not the recipe for its use.

This general focus can be applied profitably to two key areas of interpersonal attraction research—namely, studies of self-disclosure and role taking, or empathy. Such an application (as the following sections will show) creates a fundamental shift of emphasis in the ways in which relationships are viewed. Furthermore, it opens up exciting possibilities for developing our conceptualizations both of relationship development and of relationship decline.

SELF-DISCLOSURE: INFORMATION AND COMMUNICATIVE STRATEGY

This section examines some of the previous work on self-disclosure and argues that it fails to lay enough stress on the active strategic significance of self-disclosure in naturally developing relationships. Not only have researchers generally employed strangers as subjects, but even when they involve the notion of disclosure changing with a deepening relationship, they view it very much as the product of the deepening relationship rather than as a process of communication that directly drives the development of the relationship. We argue here that we must change our emphasis toward this latter view, that self-disclosure is a purposeful and strategic process of communication that is used intentionally to produce relationship development.

Previous Work on Self-Disclosure

Much of the early work on self-disclosure was stimulated by the creation of Jourard's (1958) self-disclosure questionnaire. This was made up of sixty conversational items that individuals could choose to talk about with a partner (such as their taste in music, ambitions in work, ideals of physical appearance, and so on). Subjects were asked to respond to each item by indicating the extent to which the information had been revealed to a number of target people. Variations on the questionnaire have changed the target people, the scale, and in some cases the items themselves. These cumulative changes have resulted in a mass of data dealing with the repeated disclosures of certain groups of certain people (Pederson & Higbee, 1968, review many of these studies). However, doubts remain as to the predictive validity of the questionnaire. In several studies (for example, Ehrlich & Graeven, 1971), subjects' actual observed disclosures did not correlate with their disclosure pattern as reported on the questionnaire. However, on closer examination, we see that in this experiment, as in many others, the subject is asked to actually disclose to a stranger, whereas the questionnaire assesses past disclosures to a parent or best friend. So it is hardly surprising that no correlation was found between the two measures. Indeed, where subjects are asked to say, on a similar questionnaire, what they would be prepared to disclose to a same-sex stranger, there is a high correlation with their behavior in experiments (Jourard & Resnick, 1970).

By ignoring the actual pattern of disclosure—how pieces of personal information are actually communicated between partners—these early studies were bound to find contradictory and confusing results. An individual's report of all his or her past disclosures ignores the relational perspective that is crucial to an understanding of the communicaiton of personal information. Such a perspective must concentrate not only on what both partners say to each other (and how they interpret what is said to them), but also on the goals they have in mind when they disclose. We lose much of significance if we adopt Ehrlich and Graeven's (1971) definition that "self disclosure in its operational usage refers to all verbal statements that a person makes about himself *regardless of his intent or motive for making the statement*" (p. 389, emphasis added). Our current work on self-disclosure argues that the meaning of a disclosure in a relationship is determined mainly by the intentions and goals of the discloser, and that a statement does not have any absolute level of intimacy or meaning in itself.

Altman and Taylor's (1973) model of intimacy growth led to increased emphasis on the study of both discloser and recipient. It put intimacy growth firmly onto an interpersonal level, developing along with a deepening relationship. It stressed the intimacy level of items disclosed between partners, and experimental studies linked this factor with several others which had earlier been seen as important (for example, age, sex, and timing of disclosure). However, Altman and Taylor's scale of intimacy-related topics contributed to the view that disclosures were a product of the relationship. For example, it was implied that certain topics were necessarily very intimate and therefore would not be talked about until the relationship was itself very close and intimate. Instead, we would argue that any item can be discussed in an intimate or nonintimate style, allowing it to be raised at any point in the relationship, and with the style used determining later developments in the relationship. For example, an item rated by Altman and Taylor as highly intimate may be broached by a casual and superficial disclosure early on in the relationship. The partner's reaction to this disclosure would allow the discloser to make inferences about whether or not a more intimate discussion of that topic could then be embarked on without fear of embarrassment or rejection.

Such flexibility in the level of intimacy assigned to any particular topic by a discloser is not only useful as a means of eliciting information from the other person, but also serves to define and characterize the intimacy level of their relationship. It is as if any and all topics can be talked about at a superficial level of intimacy between friends. Morton's (1978) study of the conversations of spouses and strangers clearly illustrates these processes of the trivialization and personalization of conversation topics. Thus, again, we argue for stress on studying the process of communicating personal information: on how disclosures may be used to change the direction of a relationship's development, rather than a rigid adherence to the view that disclosures (and their intimacy level) are a simple product of the stage of a relationship.

Many of the experimental studies of self-disclosure (see Cozby, 1973, for review) have centered on the investigation of the link between self-disclosure on the one hand and liking or ratings of mental health on the other. However, conflicting results have often been found, always mediated by other factors such as sex norms or the type of information revealed (Miell, Duck, & La Graipa, 1979; Archer & Burleson, 1980). Chelune (1976, 1979) has attempted to clarify the issue by stressing that people use rules to judge whether or not a particular disclosure is appropriate in a given situation and that their disclosures are not merely the product of that situation; instead, people can and must be flexible in their disclosures. Kelly (1955, p. 31) has also suggested that a lack of flexibility in dealings with others might be a major cause of psychological disorder: "From the standpoint of the psychology of personal constructs, we may define a disorder as any personal construction which is used repeatedly in spite of invalidation." Similarly, Snyder and Monson's (1975) notion of self-monitoring may be useful here, where individuals are sensitive to the interpersonal cues to appropriateness and act on them.

Again, this raises the notion of people actively monitoring and assessing their behavior (and that of their partner; Ajzen, 1977) in order to disclose in the most effective manner. Derlega and Grzelak (1979) further stress the importance of examining the intentions and beliefs held by disclosers if we are to make sense of their observable behavior. They also suggest that disclosures are seen by ordinary individuals as serving functions for the discloser—functions such as aiding in self-expression, developing a relationship, or facilitating social control. Thus, recent research is examining why we disclose and how we may use disclosure to achieve certain goals, rather than, as previously, viewing increased disclosure as simply the inevitable by-product of a deepening relationship, and therefore in some way beyond the control of the partners in the relationship.

Present and Future Work

In our view, this stress placed on the functions of self-disclosure by Derlega, Chelune, and others is a useful development in the literature. For the first time, disclosures are seen as useful, and therefore used by disclosers in, for example, directing the course of a relationship's development, rather than being merely the product of that development. Such a view necessitates a shift in focus away from the purely observable behavior of dyads of strangers to an examination of the behavior and intentions of disclosers as they use their disclosures to direct the course of a relationship's development.

This requires two developments in our research. First, we must place greater emphasis on studying actual disclosures as a naturally occurring relationship develops, unlike the earlier experimental work using strangers, and often only looking at one person's disclosures. If we see self-disclosure as serving such functions as developing a relationship and aiding in social control, we must examine it in these contexts, as the nature of the disclosures in each may be fundamentally different. For example, the same piece of personal information may be revealed at different levels of intimacy, or in varying manners (as a joke, seriously, tentatively, and so on) in order to achieve the particular aim the discloser has in mind. A second requirement is that we shift our emphasis from the study of the individual and his or her disclosures toward that of the dyad. It is the communication of information that is important, and thus it is both partners' disclosures—and their intentions about these disclosures—that should be the focal concerns of our future research.

At Lancaster, we are currently examining the aims and intentions of partners in developing their relationship (Miell & Duck, 1983) and the strategies of disclosure that they employ to achieve thes ends. Here, strategic disclosures are seen to be those that are deliberately used by an individual to, for example, create a particular impression of themselves to another, or to influence later disclosures by a partner. We believe (Miell & Duck, 1983) that there is a hierarchy of such aims and strategies, some being more consciously used than others, some being more appropriately applied to long-term planning for a relationship, and others relating to particular conversations or even the choice of phrasing of a disclosure.

This view, then, is of partners as active planners in their relationship, using the revelation of personal information as a means of effecting desired changes in the relationship's development. It also lays particular stress on the processes of communication and negotiation between the partners (who also, in this model, require strategies and counterstrategies to deal with each other's strategies). As a result, our methods revolve around the analysis of partners' actual disclosures over the course of their naturally occurring relationship. Both partners are also asked to analyze examples of their disclosures, allowing us to compare their views of the same conversation, and indeed of particular disclosures. They are then asked to discuss their plans for the relationship's future and to comment retrospectively on its past. A reinterpretation of the history of the relationship in light of what each partner now knows to have happened is as evident in these studies as it has been in several other studies looking at naturally occurring relationships (Duck & Miell, 1981, 1982, in press). Particular incidents, and especially conversations, appear to take on greater or lesser significance depending on the later course of development of the relationship. Prospective views of the future development of the relationship tend to revolve around the partners' plans for the relationship—plans which then direct the general approach that one person adopts when conversing with his or her partner (Miell & Duck, 1983). During these conversations, any possibility that a "taboo"topic may be discussed will be avoided by a strategic move to disclose an alternative, safer piece of information. Subjects were asked to comment on what was happening in a conversation with a friend they had just recorded on videotape. One subject commented:

> Now, there I nearly went further than I intended to so I stopped, that's why I hesitated, I started to say something then I thought "no, that's something that I don't think even my best friend knows," so therefore I wasn't about to commit myself on that, so I just stopped and fished around for something else that would lead into another topic, that's why there was that gap. We went on to building a wardrobe—that was it, yes. Well, it was a problem connected with my boyfriend, which brought my boyfriend into mind, which brought the wardrobe into mind; it was an association of ideas really. I accept it as a superficial relationship and it comes across on this tape as such because every time it gets too personal the subject's changed—either by Margaret or by myself. I think it comes across as my having vaguely lost the drift of the conversation, because I struck myself as being a bit vague when I was doing it, but it was only because I was obviously thinking "God, I musn't say that I've got to say something else." I would have thought that Margaret would have accepted it as a general sort of "she's got something else on her mind that she wants to say"; she may well not have noticed at all in fact.

This shows how the subject's overall plan for the relationship determined which topic was currently taboo in the relationship (in this case, problems to do with a boyfriend), but particular strategies are what allow the person to actually sidestep the issue successfully.

Through the use of interviews and analysis of videotaped conversations between friends, both by the friends themselves and by an external observer, the complicated network of aims and intentions surrounding the use of disclosures in developing relationships is beginning to be understood (Miell & Duck, 1983). It will lead to a more fruitful view of relationship development and decline if we adopt this view of the importance of the process of communication between friends as evidenced in the disclosure and exchange of personal information. Certainly, work at Lancaster (Duck & Miell, 1982, in press) and Dundee (Emler & Fisher, n.d.) using diary accounts of everyday interactions suggests that personal disclosures constitute a large proportion of the general information exchanged in conversations between friends and acquaintances.

EMPATHY AS A COMMUNICATIVE PROCESS
OF RELATIONSHIP DEVELOPMENT

In this section, previous work on empathy is shown to have neglected (a) consideration of empathy as a significant component of models of relationship development, and (b) explication of the communicative processes by which empathy evolves in dyadic relationships. New research is reported which combines these relational and communicative approaches and provides support for a new theory of empathy as a communicative process of relationship development. In addition, a novel methodology, known as "role exchange," is described as a means for generating data on empathic communication in developing relationships.

Previous Work on Empathy

Since its introduction by Lipps at the turn of the century, the term "empathy" has undergone numerous and varied interpretations. What remains is a bewildering diversity of conceptual, methodological, and empirical contributions to the field. This field of inquiry ought to be central in a psychology of relationship growth and decline (insofar as it comprises the issues of how persons may come to understand, and misunderstand, each other). Yet concepts of empathy, and the empirical results that have accumulated, have never been integrated within the mainstream of theory and research in relationship development. We suggest that this is another particular instance of the regrettable lack of attention to processes of relationship development. Instead, as with the previous literature on "attraction," research on empathy has identified in piecemeal fashion, typically through correlational studies, a plethora of traits, abilities, and behaviors of individuals that have been defined, ad hoc, as "empathic." For example, Stotland (1969) found physiological correlates of empathy; Truax's empathy scale identified individuals with effective counseling skills (Truax & Carkhuff, 1967); Aderman and Berkowitz (1970) measured individuals' changes in mood in response to exposure to emotionally arousing stimuli; Borke (1971) studied children's ability to infer emotions from stereotyped facial expressions as a function of ontogenetic development; Mehrabian and Epstein's (1972) scale measured individual differences in empathy which correlated positively with altruism and negatively with aggression; and Dymond's (1949) operational definition of empathy was the ability to predict another's questionnaire responses.

Although this last example (chronologically one of the first detailed empirical studies) approaches a rational measure, the overwhelming majority of research over several decades (of which this list is just a small part) has proceeded without reference to dyadic relationships, far less their development, and indeed mostly without any integrating conceptual framework. Thus, Hornblow's (1980) review of the literature concudes: "There is no agreed-on definition of "empathy," nor is there any generally accepted theoretical model for understanding empathic processes" (p. 25). Henceforth, he argues that "empathy researchers will need to place less emphasis than in the past on empathy as a global attribute and give more attention to its situational determinants. . . . to identify ways in which contextual factors facilitate or inhibit empathic interaction" (p. 26).

In our view, two specific developments are called for. First, attention should switch from the individual to the dyadic relationship, for it is here that empathy develops. It is almost by definition *not* a property of individuals but of relationships between individuals. One major advance in this direction has been the work of Laing, Phillipson, and Lee (1966), pursuing Dymond's (1949) initiative, who see empathy not simply as the knowledge of how one's partner feels, but, at higher "levels" of empathy, the knowledge of one's partner's knowledge of one's knowledge of this or that issue in what, logically, is an infinite recursion of understandings (and/or misunderstandings).

In empirical investigations, using their Interpersonal Perception Method (see also Bryant, 1974; Knudson, Sommers and Golding, 1980), qualitatively distinct kinds of understanding and misunderstanding can arise for all issues of relevance to the partners concerned, corresponding to the different levels in this proposed recursive structure. However, Laing et al. (1966) while recognizing that lack of understanding at the levels of metaperspectives and metametaperspectives (and, we suggest, beyond) may result from failures in communication within the dyadic relationship (leading, for example, to failures to realize misunderstanding of a disagreement), do not address empirically the second major question: How do empathic relationships and, in particular, these various recursive levels of empathy, develop through dyadic interactions?

Little work has been concerned with communication as a process in developing empathic relationships, and still less has considered how recursively structured empathic information might be communicated. Most work on relational communication generally has been undertaken within the theoretical frameworks of self-disclosure and has focused predominantly on verbal behavior. It is clear, however, from an inspection of sentences of the kind constructed by Laing et al. in describing the contents of these higher levels of understanding (for example, husband thinks wife doesn't think husband thinks wife thinks . . .) that direct verbal expression is awkward and unfamilar. These higher-level sentences do not appear in our everyday communication either because the empathic information they contain is of no significance or because these higher levels are communicated by some other means. The possibility that the same information is communicated nonverbally has not previously been explored, despite a well-established tradition of research into nonverbal behavior (for example, Argyle, 1969). Both the relational and communicative approaches are combined within the perspective adopted by the present authors. This perspective assesses the degree of similarity and empathy by analyzing both partners' communications to each other of significant aspects of their perceptions of self and other. The growth of empathy in a dyadic relationship is seen here as a process in which both partners communicate their understandings of each other across all relevant levels in the recursive structure suggested by Laing et al.'s (1966) model.

Intimation Sequences

Dyadic interactions comprise, naturally enough, both verbal and nonverbal behavior, yet rarely have researchers considered the interactions between these two channels of communication, less often still from the perspective of relationship development (but see Miller & Parks, 1982). Typically, research on verbal behavior in developing relationships has continued almost as if nonverbal behavior did not occur. Likewise, research on nonverbal behavior as an indicator of growth in dyads has tended to see communication of this kind as a topic independent of verbal communiction (studies of gaze, pupil dilation, interpersonal distance, and so on as a function of intimacy in a relationship; see Argyle, 1969). Yet nonverbal channels have a special property that would suggest a valuable contribution to the communication of recursively structured empathic information. That is, they can operate in conjunction with verbal communication. Indeed, nonverbal signals often have no other meaning than that provided by the context of ongoing verbal communication (Watzlawick, Beavin, & Jackson, 1968).

For example, a nod of the head has little meaning in itself (except that it probably conveys affirmation or agreement), whereas it immediately takes on quite precise and unique meaning when interpreted in the context of an accompanying verbal disclosure (typically, a disclosure by the other person). It is this "metacommunicative" property (Watzlawick et al., 1968) of nonverbal signals that is explored here as the vehicle for the communication of recursively structured empathic information.

It is proposed that a sequence of the appropriate nonverbal signals, occurring in the context of a verbal disclosure, may convey recursively structured empathic representations as complex as "husband is aware that wife is surprised that husband had not realized wife's feelings about a given topic." Suppose in a dyadic interaction in which all requisite channels are optimally available that wife indicates her feelings about a given topic through a direct verbal disclosure. Husband may then look surprised, for only now does he become aware of his wife's feelings on this topic. Wife notices husband's surprised reaction and looks surprised herself, for she had always assumed that her husband understood her feelings on this topic. Husband notices his wife's surprised reaction and smiles in realizing her new awareness. This sequence of signals, in which both partners intimate to each other new levels in the evolving recursive structure of empathic representations that predicate the given topic, is termed an intimation sequence.

It is proposed that the development of empathic relationships is a process involving self-disclosure and intimation sequences, not separately, but essentially bound together in every face-to-face dyadic encounter. As each new level of one's representations arises, the possibility exists that an accompanying behavioral event will intimate this representation to the partner and enable the partner to construct a new level of empathic representation. Provided that optimal interaction can occur, the couple may progress rapidly through the recursive structure until all relevant levels have been fully established and maximum empathy obtained as far as both partners are concerned. On the other hand, failure to achieve the higher levels of empathy may ensue where communication is significantly constrained. This could happen, for example, through inability or disinclination to attend to intimation signals and/or verbal disclosures, such as might arise in cognitively immature or inept populations (such as children or clinical patients), or as a result of deliberate or involuntarily imposed constraints on the availability of information.

Role Exchange

To test the main hypothesis that intimation sequences are a behavioral manifestation of the communicative process of constructing recursive empathic representations, a new methodological technique, known as role exchange, has been developed for generating data on couples' empathic behavior in dyadic interactions (Miell, 1980, 1981). In role exchange, two close friends are asked to "become each other," as it were, in everything they say during a conversation between them on a given discussion topic. Their interaction is recorded on videotape for subsequent analysis. The role exchange task is designed to encourage the construction and communication of all relevent levels of empathic representation so that the process of empathy may be observed. Detailed analysis of a large sample of videotaped role exchange has revealed sequences of intimation signals (nods, raised eyebrows, smiles, grins, and glances) that vary, in the complexity of recursive levels they imply, from many two-step sequences (for example, husband's nod followed by wife's smile) through, with decreasing frequency of occurrence, six-step sequences.

Empathy and Relationship Development

Studies using the role exchange technique (Miell, 1981) have investigated empathic communication in couples of varying relationship durations. If, as suggested by the model described above, relationship development is characterized not only by an increase in the range of topics on which the couple have disclosed to each other (Altman & Taylor, 1973) but also by the extent to which the potential recursive structure of empathic representations predicating that topic has been explored, then one would expect role exchange behavior to show a developmental trend in relationships. Specifically, it was hypothesized that if role exchange discussion topic was held constant, length of intimation sequence should vary inversely with relationship duration of the role exchange couple. Thus, couples who had fully explored all relevant levels of representation prior to experimental role exchange would make (and intimate) fewer significant empathic discoveries during role exchange than couples of shorter relationship duration.

An operational test of this hypothesis required a modification of the measures of the dependent variable, since not all intimation sequences during role exchange were attributable to the role exchange manipulation. Instead, a global measure was introduced. It was hypothesized that if the videotaped role exchanges were shown to observers whose task was to decide whether or not role exchange was occurring, they would correctly identify role exchange only for those (shorter duration) couples who were seen, by their otherwise inappropriately long intimation sequences, to be making significant empathic discoveries as a result of the role exchange manipulation. By contrast (longer duration) couples whose role exchange behavior would have no more intimation than a normal interaction on the same topic (since there were no significant discoveries to be made as a result of role exchange) would be confused with normal (non-role-exchange) interactions. The results of this "intimation game" test of the hypothesis were strongly supported.

Relationship development may thus be seen as a communicative process in which verbal self-disclosure and nonverbal intimation sequences operate to bring about both breadth and depth in the recursive structure of empathic representations. As already suggested, constraints on opportunities for effective disclosure and intimation in face-to-face dyadic interactions would inevitably impair the evolution of high-level empathy and thereby contribute to a gradual process of relationship decline. More optimistically, it should be recognized that a misunderstanding at one level may, through effective empathic communication, be resolved by first achieving a higher-level awareness of this lower misunderstanding.

COMMUNICATIVE STRATEGIES IN RELATIONSHIP DECLINE

Although the discussions in the foregoing sections have been largely focused on the processes of making relationships develop, much of what has been said could be stood on its head and made to apply to relationship decline. In this concluding section, we will relate some of the above to a recently developed map of relationship breakdown and dissolution (Duck, 1982a), and illustrate applications of the foregoing principles. A specific consideration of the means by which people choose communicative strategies to effect the dissolution of a relationship can be found in Miller and Parks (1982).

As can be seen from Figure 11.1, there are a number of contiguous but different phases to the breakdown and dissolution of relationships. The later phases concern the communication (to out-siders) of information about the ending of a relationship. Of greater concern in the present context are the breakdown, intrapsychic, and dyadic phases. In each of these, the communicative skills and behaviors discussed above have a direct bearing on the success, growth, or decline of the relationship.

It has been our contention that communication about certain aspects of oneself is a more significant influence on relationship growth than are those aspects themselves. We have stressed the strategic nature of such communication and have laid emphasis on its interweavig with complex levels of understanding in rela-tionships. Clearly, two individuals who do not communicate satisfactorily, be they ever so similar, will be unable to establish valuable patterns of interaction that are facilitative of worthwhile and enjoyable relationship growth. Research should focus less on individuals who remain unaware of the strategic needs of their part-ner, and more on those who cannot conduct the joint strategic negotiations that do occur. Equally, we should attend less to those who are unable to achieve proper and appropriate levels of understanding and focus on those who do not communicate their level of understanding appropriately. However, unintentional failure to communicate is not the only means to achieve the dissolution of a relationship. Self-disclosure can itself be used strategically in relationships, both to facilitate their growth, as we have suggested above, or to create their dissolution, as Baxter (1982) has shown.

There is still useful work to be done on the communicative deficits or disappointments that cause problems during the dif-ferent phases of dissolution. At the breakdown phase, where one or both partners is/are experiencing dissatisfaction with the rela-tionship (while perhaps retaining a positive feeling toward their partner), it is likely that the conduct of communication in the rela-tionship is at fault. For instance, partners may be failing to read another empathically, as discussed earlier. In the intrapsychic phase, on the other hand, it is more likely that partners "are not communicating" in the familiar, colloquial sense. Their feelings, emotions, and private needs are perhaps inadequately expressed, and strategic self-disclosure of the kind discussed in an earlier sec-tion is required to put the balance right. According to Duck's (1982a) model, the intrapsychic phase is founded on a sense of grievance and distress at the partner's insensitivity or incapacity to fulfill one's needs adequately. Whether the balance is best restored by the methods discussed above is an open empirical

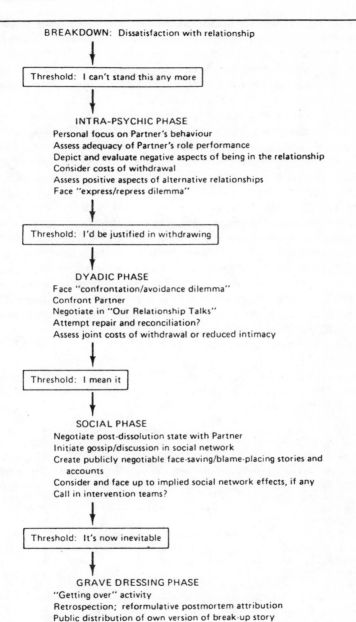

BREAKDOWN: Dissatisfaction with relationship

↓

Threshold: I can't stand this any more

↓

INTRA-PSYCHIC PHASE
Personal focus on Partner's behaviour
Assess adequacy of Partner's role performance
Depict and evaluate negative aspects of being in the relationship
Consider costs of withdrawal
Assess positive aspects of alternative relationships
Face "express/repress dilemma"

↓

Threshold: I'd be justified in withdrawing

↓

DYADIC PHASE
Face "confrontation/avoidance dilemma"
Confront Partner
Negotiate in "Our Relationship Talks"
Attempt repair and reconciliation?
Assess joint costs of withdrawal or reduced intimacy

↓

Threshold: I mean it

↓

SOCIAL PHASE
Negotiate post-dissolution state with Partner
Initiate gossip/discussion in social network
Create publicly negotiable face-saving/blame-placing stories and
   accounts
Consider and face up to implied social network effects, if any
Call in intervention teams?

↓

Threshold: It's now inevitable

↓

GRAVE DRESSING PHASE
"Getting over" activity
Retrospection; reformulative postmortem attribution
Public distribution of own version of break-up story

SOURCE: Reproduced with permission from S. W. Duck, **Personal Relationships 4: Dissolving Personal Relationships**, page 16. Copyright © Academic Press, Inc. (London) Ltd.

Figure 11.1: Sketch of the Main Phases of Dissovling Personal Relationships

question. However, our argument is that a certain type of communication is required to set things right, and that it is a type that
differs from that needed to correct the balance at different phases
of relationship dissolution. When the partners enter the dyadic
phase, the main issue is the selection of a communication strategy
that best deals with the negative feelings that, at that time, provide the only warmth in the relationship. Baxter (1982) and Miller
and Parks (1982) have given extended discussion to the issues here
and lay out some intriguing possibilities for further research.

Thoroughgoing analyses of the causes, course, and prevention
or facilitation of relationship dissolution are only just begining to
get the major theoretical and research attention that they merit
(Duck, 1981, 1982b, 1984). Social behavior in relationships is not
an epiphenomenon nor a mere reflection of some energetic motive
force within a person's cognitive structure. Communicative social
behavior in relationships is a social tool. People have social in-
tents, social and relational goals, strategies for achieving or constraining relational growth, and social mechanisms for managing
their relationships. As we have tried to show in the present chapter,
communicative behavior is a means by which these are achieved,
made real, and brought to life.

REFERENCES

Aderman, P., & Berkowitz, L. (1970). Oberservational set, empathy and helping.
    Journal of Personality and Social Psychology, 14, 141-148.
Ajzen, I. (1977). Information processing approaches to interpersonal attraction. In
    S. W. Duck (Ed.), *Theory and practice in interpersonal attraction.* London:
    Academic Press.
Altman, I., & Taylor, D. A. (1973). *Social penetration: The development of interper-
    sonal relationships.* New York: Holt, Rinehart & Winston.
Archer, R. L. & Burleson, J. A. (1980). The effects of timing of self-disclosure on
    attraction and reciprocity. *Journal of Personality and Social Psychology, 38,*
    120-130.
Argyle, M. (1969). *Social interaction.* London: Methuen.
Baxter, L. A. (1982). *Disengagement as a process rather than an event.* Paper
    presented to International Conference on Personal Relationships, Madison,
    Wisconsin, July.
Berger, C. R., & Bradac, J. J. (1982). *Language and social knowledge: Uncertainty
    in interpersonal relations.* London: Edward Arnold.
Berscheid, E., & Walster, E. (1974). "Physical attractiveness." In L. Berkowits (Ed.),
    Advances in experimental social psychology: Vol. 7. New York: Academic Press.
Borke, H. (1971). Interpersonal perception of young children: Egocentrism or em-
    pathy? *Developmental Psychology, 5,* 263-269.
Bryant, B. K. (1974). Locus of control related to teacher child interpersonal ex-
    periences. *Child Development, 45,* 157-164.

Byrne, D. (1971). *The attraction paradigm.* New York: Academic Press.

Byrne, D., Ervin, C. R. & Lamberth, J. (1970). Continuity between the experimental study of attraction and real-life computer dating. *Journal of Personality and Social Psychology, 16,* 157-165.

Chelune, G. J. (1976). A multidimensional look at sex and target differences in disclosure. *Psychological Reports, 39,* 259-263.

Chelune, G. J. (1979). Measuring openness in interpersonal communication. In G. J. Chelune et al. (Eds.), *Self disclosure.* San Francisco: Jossey-Bass.

Cozby, P. C. (1973). Self-disclosure: A literature review. *Psychological Bulletin, 79,* 73-91.

Derlega, V., & Grzelak, J. (1979). Appropriateness of self-disclosure. In G. J. Chelune et al. (Eds.), *Self disclosure.* San Francisco: Jossey-Bass.

Duck, S. W. (1973). *Personal relationships and personal constructs: A study of friendship formation.* London: John Wiley.

Duck, S. W. (1977). *The study of acquaintance.* Farnborough: Teakfields.

Duck, S. W. (1981). Toward a research map for the study of relationship breakdown. In S. W. Duck & R. Gilmour (Eds.), *Personal relationships 3: Personal relationships in disorder.* London: Academic Press.

Duck, S. W. (1982a). A typography of relational disengagement and dissolution. In S. W. Duck (Ed.), *Personal relationships 4: Dissolving personal relationships.* London: Academic Press.

Duck, S. W. (Ed.). (1982b). *Personal relationships 4: Dissolving personal relationships.* London: Academic Press.

Duck, S. W. (Ed.). (1984). *Personal relationships 5: Repairing personal relationships.* London: Academic Press.

Duck, S. W., & Miell, D. E. (1981). *Charting the development of relationships.* Paper presented at conference on Long Term Relationships, Oxford University, November.

Duck, S. W., & Miell, D. E. (1982). *Charting the development of personal relationships.* Paper presented to International Conference on Personal Relationships, Madison, Wisconsin, July.

Duck, S. W., & Miell, D. E. (in press). Toward a comprehension of friendship development and breakdown. In H. Tajfel, C. Fraser, & J. Jaspars (Eds.), *The social dimension: European perspectives on social psychology.* Cambridge: Cambridge University Press.

Duck, S. W., & Sants, H. K. A. (1983). On the origin of the specious: Are personal relationships really interpersonal states? *Journal of Social and Clinical Psychology, 1.* 15-24.

Dymond, R. F. (1949). A scale for the measurement of empathic ability. *Journal of Consulting Psychology, 13,* 127-133.

Ehrlich, H., & Graeven, D. (1971). Reciprocal self-disclosure in a dyad. *Journal of Experimental Social Psychology, 7,* 389-400.

Emler, N., & Fisher, S. (n.d.). *Social pariticipation as personal information exchange.* Unpublished manuscript, University of Dundee, Scotland.

Hinde, R. A. (1979). *Towards understanding relationships.* London: Academic Press.

Hinde, R. A. (1981). The bases of a science of personal relationships. In S. W. Duck & R. Gilmour (Eds.), *Personal relationships 1: Studying personal relationships.* London: Academic Press.

Hornblow, A. R. (1980). The study of empathy. *New Zealand Psychologist, 9,* 19-28.

Jourard, S. M. (1958). The study of self-disclosure. *Scientific American, 198,* 77-82.

Jourard, S. M., & Resnick, J. (1970). Some effects of self-disclosure among college women. *Journal of Humanistic Psychology, 10,* 84-93.

Kelly, G. A. (1955). The psychology of personal constructs. New York: W. W. Norton.

Kerckhoff, A. C., & Davis, K. E. (1962). Value consensus and need complementarity in mate selection. American Sociological Review, 27, 295-303.

Knudson, R. M., Sommers, A. A., & Golding, S. L. (1980). Interpersonal perception and mode of resolution in marital conflict. Journal of Personality and Social Psychology, 38, 751-763.

Laing, R. D., Phillipson, H., & Lee, A. R. (1966). Interpersonal perception. London: Tavistock.

Levinger, G. (1980). Toward the analysis of close relationships. Journal of Experimental Social Psychology, 16, 510-544.

Mehrabian, A., & Epstein, N. (1972). A measure of emotional empathy. Journal of Personality, 40, 525-543.

Miell, D. E., & Duck, S. W. (1983). Strategies in developing close relationships, Paper presented to second international conference on Language and Social Behaviour, Bristol, July.

Miell, D. E., Duck, S. W., & La Gaipa, J. J. (1979). Interactive effects of sex and timing on self disclosure. British Journal of Social and Clinical Psychology, 18, 355-362.

Miell, D. K. (1980). The intimation game. Paper presented at the Annual Conference of the Social Psychology Section of the British Psychological Society, Guildford, September.

Miell, D. K. (1981). Recursive interpersonal cognition: Empathic representation, communication and role exchange. Unpublished thesis, University of Lancaster.

Miller, G. R., & Parks, M. (1982). Communication in dissolving relationships. In S. W. Duck (Ed.), Personal relationships 4: Dissolving personal relationships. London: Academic Press.

Morton, T. (1978). Intimacy and reciprocity of exchange: A comparison of spouses and strangers. Journal of Personality and Social Psychology, 36, 72-81.

Morton, T., & Douglas, M. (1981). Growth of relationships. In S. W. Duck & R. Gilmour (Eds.), Personal relationships 2: Developing personal relationships. London: Academic Press.

Pederson, D. M., & Higbee, K. L. (1968). Self-disclosure and relationship to the target person. Merrill Palmer Quarterly, 15, 213-220.

Perrin, F.A.C. (1921). Physical attractiveness and repulsiveness. Journal of Experimental Psychology, 4, 203-217.

Snyder, M., & Monson, T. C. (1975). Persons, situations and the control of social behaviour. Journal of Personality and Social Psychology, 32, 637-644.

Stotland, E. (1969). Exploratory investigations of empathy. In L. Berkowitz (Ed.), Advances in experimental social psychology: Vol. 4. New York: Academic Press.

Traux, C. B., & Carkhuff, R. R. (1967). Toward effective counselling and psychotherapy. Chicago: Aldine.

Watzlawick, P., Beavin, J. H., & Jackson, D. D. (1968). Pragmatics of human communication. London: Faber.

Figure 11.1: Sketch of the Main Phases of Dissolving Personal Relationships
SOURCE: Reproduced with permission from S. W. Duck, Personal Relationships 4: Dissolving Personal Relationships, page 16. Copyright © Academic Press, Inc. (London) Ltd.

AFTERWORD

Recent Developments in Social Cognition and Interpersonal Behavior:
A Consolidation and Extension

The chapters in this book represent some of the most important recent trends in the burgeoning research area of social cognition. My purpose in this brief afterword is to position the present work within the context of the more general developments in research on social cognition. I believe the volume makes an important general contribution to our understanding of social cognition and interpersonal behavior while advancing on several specific problems that are focal to the individual chapters.

Social cognition has, of course, been with us as an area of study for quite a long time. Certainly for those interested in social cognition and communication and their development, as is the case of several contributors to the present volume, research dates back more than fifty years to work stimulated by the seminal contributions of Piaget and Mead. Concern with social intelligence, social sensitivity, and the accuracy of social judgments likewise dates from the pre-World War II period, and the silver anniversary has already passed for the publication of Heider's (1958) *The Psychology of Interpersonal Relations.* Tagiuri and Petrullo's (1959) edited volume, which marked a shift in social perception research toward greater concern with interpersonal behavior, is also at its 25th publication anniversary.

It is evident to all of us who work in the area, however, that something has happened in the past decade. Social cognition research has shifted from a set of loosely related topics of study to what is now becoming widely recognized as an important, increasingly integrated area of research. This new status reflects, in part, the general ascendancy of cognitive psychological models, but it represents much more than changes in particular subfields in psychology. Indeed, the study of social cognition appears to be coalescing at the juncture of at least four fields of study: social and personality psychology, child development, cognitive psychology, and speech communication. In addition, contributions to the increasingly consolidated work in the area are being made by anthropologists, linguists, and sociologists.

This trend toward interdisciplinary consolidation is reflected in the initiation of such interdisciplinary journals as *Social Cognition,* the *Journal of Social and Personal Relationships,* and *Social Psychology and Language,* and in a host of recent volumes surveying current trends in theory and research (for example, Cantor & Kihlstrom, 1981; Forgas, 1981; Harvey, 1981; Hastorf & Isen, 1982; Higgins, Herman, & Zanna, 1981; Higgins, Ruhle, & Hartup, 1982; Ickes & Knowles, 1982; Roloff & Berger, 1982; Serafica, 1982). These publications have established the importance of recent theoretical developments for traditional lines if work and corrected some of the biases that have reflected narrow disciplinary concerns. The present volume follows in this vein, though it is less self-consciously directed toward research consolidation and assessment than have been several of its recent predecessors. This difference reflects, I think, the accomplishment of interdisciplinary consolidation with a consequent turn toward work from a variety of perspectives on matters of common interest to a group of researchers. In the following pages, I want to briefly highlight several themes in the present volume that reflect this emergent consolidation of social cognition as an area of interdisciplinary study.

If there is a traditional core to work on social cognition, it is in research tied, however loosely, to Asch's (1946) classic study of impression organization. In that study, subjects were provided with trait information describing a target, and the patterns of inferences drawn from differing sets of information were the focus of analysis. The theoretical and methodological architecture of this study were given numerous twists and turns in the ensuing thirty years, but through all the permutations there were some constancies. Among these were the following: First, there has been a tendency to study social cognition from the standpoint of the observer and, therefore, to divorce the study of social cognition from the structure of action (see, for contrast, the phenomenological sociology of Schutz, 1967; Schutz & Luckmann, 1973; or Heider, 1958). Moreover, when social cognition was studied as a foundation of interpersonal behavior, it tended either to be through the use of a broad and general concept like egocentrism or, in the 1970s, through a narrowed concern with the attribution of actional causality.

Second, social cognition also tended to be treated purely in terms of a cognitive/social inference model. The social inference process was conceived as reflecting the content and organization of the perceiver's cognitive system (or implicit personality theory). Thus, in Piaget's terms, the dominant model represented cognitive processes as highly assimilatory. Models emphasizing accommodation to target characteristics in social perception tended to be so methodologically constrained (such as Anderson, 1962) or concerned with the issue of judgmental accuracy (such as Cline, 1964) that they had little impact beyond narrow subfields.

Third, there was a strong tendency to study social cognition apart from social and interactional contexts, and hence to restrict the focus of social cognition to perceptual and judgmental processes. Until quite recently, there was little systematic concern with information seeking, memory and informational retrieval, or the role of social context in social cognition research.

During the past decade, many of the elements of the traditional paradigm have been supplanted. The most important developments doubtlessly have been the concern given attribution processes by social psychologists, the emergence of sustained lines of research on social cognitive and communicative development, and the rise of cognitive science as a major force in the psychological and behavioral sciences. The turn away from general processes of impression formation has been supplanted by concern with processes involved in the construal of particular events, the organization of social knowledge across times and contexts, and the relationship of social cognition to social behavior. No new synthesis has yet emerged, but these themes in recent work on social cognition, most of which are clearly reflected in the present volume, clearly involve a significant departure from the questions and models being vigorously pursued little more than a decade ago.

The present volume includes some very useful efforts to integrate and extend the major concepts that have been used in the effort to link social cognition and behavior: perspective taking (Barnett) and schema (Housel). However, there is also a less overt contribution to our general theoretical understanding of the social cognition/interpersonal behavior relationship present in other chapters. Several contributors shift from reliance on global concepts like perspective taking to more focused aspects of social cognition that have clearer relevance to particular aspects of social action (for example, Burleson's analysis of the role of abstract modes of construing persons in comforting communication, or Forbes and Lubin's analysis of conceptions of social influence and persuasive tactics). Indeed, most of the chapters in the present volume reflect a much greater concern with interpersonal behavior (especially aspects of communicative and prosocial behavior) than was characteristic of work in the area of social cognition even five years ago.

This concern with finding concepts that can address the role of social cognition in communication and interpersonal behavior has led to an increased emphasis on frameworks relevant to understanding situated perceptions and action, and particularly to a major emerging concern with intentions, plans, and goal structures. Particularly significant moves in this direction are McCann and Higgins's reinterpretation of a host of individual differences as reflecting variations in typical goals, Eisenberg and Silbereisen's general action-theoretic framework, Burleson's analysis of comforting communication, and Duck, Miell, and Miell's analysis of relationship development. Much of the work in the present volume, along with other recent research, suggests that understanding the role of social cognition in behavioral production will require multiple concepts. It is likely that social cognition plays several roles in interpersonal conduct, including perhaps a role in the generation of the intentions that produce an action (see McCann & Higgins, this volume; also see O'Keefe & Delia, 1982, in press).

Just as contributors to the present volume have begun to address the social cognition/interpersonal behavior relationship at a much more refined level than traditionally has been the case, several contributors also reflect the growing awareness that interpersonal behavior cannot be understood simply by reference to social cognitive processes, no matter how refined the conceptualization and measurement. For example, Burleson nicely calls attention to an array of processes that at a minimum must be included in developing a model of comforting behavior. In a different but equally important vein, contributions such as that of Rubin and Borwick remind us that understanding the social cognition-behavior relationship is only part of the story, for social cognition and behavior have effects and social consequences that must be considered. Moreover, several of the contributions recognize that social cognition and social interaction are reciprocally influential and that the study of these processes can usefully take interaction itself as a focus (note the contributions of Oden, Wheeler, and Herzberger; and Kraut and Lewis).

There is also reflected in the present book a major, and I think very important, shift in how social cognition is thought about and studied from the interpretive or perceptual (rather than behavioral) side. It is a remarkable commentary on the magnitude of change in the field that the majority of chapters in the present volume are concerned with some aspect of behavioral production, rather than with social cognition or social perception per se. A decade ago, save for work in attribution theory, social cognition research was dominated by the assimilation-centered concern with social perception and impression formation. The shift to schema theories and to discourse processing in cognitive psychology has led to a shift from structural theories reflecting concern with implicit personality theories to cognitive processing models stressing attention, interpretation, storage, and retrieval processes. These changes have tempered somewhat the traditional emphasis in social cognition research on processes of cognitive assimilation and have cleared the ground for a move toward more contextual and accommodation-centered models of social cognition. A shift in this direction has been advocated by theorists in social psychology (Crockett, 1977; Higgins, McCann, & Fondacaro, 1982), communication (O'Keefe & Delia, 1982), and developmental psychology (Chandler, 1982; Damon, 1981). Chandler's argument for developing frameworks that genuinely permit consideration of both processes of assimilation and accommodation seems to me particularly important. His argument is as relevant to those concerned with the study of cognition in context as to those concerned with developmental cognitive change.

For those concerned with such traditional topics as person perception and impression formation, the problem lies in developing frameworks that represent social cognition both as a cognitive process (involving attention, interpretation, and so on) and as a social process (involving information seeking, hypothesis testing, and so on). The growth of social knowledge (like developmental cognitive change) involves both cognitive and social processes operating in concrete contexts over time. In the present volume, the contributions of Rubin and Borwick and, particularly, O'Keefe reflect the movement toward such a contextual model of social cognition.

One final topic also merits comment: methodology. Traditional social cognition research was dominated by the methods of psychometrics and experimental social psychology. These traditions of work continue to be important and are represented in the present volume by Kraut and Lewis's nicely executed analysis of feedback in communication. The emerging interdisciplinary field of social cognition and interpersonal behavior, however, is characterized by methodological pluralism. Recent years have witnessed a broadening of the modes of analysis employed in social cognition and communication research. This broadening of the range of readily employed methods has been extended, in particular, to embrace observational and interaction analytic methods, in-depth interviewing, and the content and structural analysis of interview protocols and behavior. These developments in methodology reflect the growing influence of developmental researchers and communication researchers in the study of social cognitive processes. Researchers in these area have been faced with the problem of understanding the interrelation of interpretive and behavioral processes. Moreover, when children are studied or when individual differences are considered, "typical" adult processes (whatever they might be) cannot be assumed. Indeed, in research with children, both social cognition (contrast Barenboim, 1977, 1981) and communication research (Asher, 1979; Burke & Clark, 1982) using highly constraining methods have been found to offer less valid and useful information than the open-ended methods that place the burden of coding on the researcher. At least in some circumstances the same is true in studying social cognition and communication in adulthood (see Delia, O'Keefe, & O'Keefe, 1982; O'Keefe & Sypher, 1981). In the present volume, the use of a range of content and interaction-analytic modes of analysis is displayed in the contributions of Burleson; Forbes and Lubin; Oden, Wheeler, and Herzberger; Rubin and Borwick; Duck, Miell, and Miell; and O'Keefe.

The expansion of the methodological horizon of social cognition and communication research may ultimately prove to be the most important result of interdisciplinary consolidation in the area. Compared to mainstream investigations of even a decade ago, much contemporary research has a greater descriptive richness that can inform the understanding of theorists and researchers in diverse camps. In addition, the presence of active researchers committed to experimental and traditional psychometric analysis serves as a constant reminder that alternative methods need to be employed with rigor and in a fashion that contributes to a general, rather than particularistic, understanding of social cognitive processes. Because of this complementary structure, the interdisciplinary study of social cognitive and interpersonal behavior offers the genuine promise of achieving a synthesis that is at once conceptually relevant, descriptively grounded, and methodologically rigorous.

REFERENCES

Anderson, N. H. (1962). Application of an additive model to impression formation. *Science*, 138, 817-818.

Asch, S. E. (1946). Forming impressions of personality. *Journal of Abnormal and Social Psychology*, 41, 258-90.

Asher, S. R. (1979). Referential communication. In G. J. Whitehurst & B. Z. Zimmerman (Eds.), *The functions of language and cognition.* New York: Academic Press.

Barenboim, C. (1977). Developmental changes in interpersonal cognitive systems from middle childhood to adolescence. *Child Development, 48,* 1467-1474.

Barenboim, C. (1981). The development of person perception in childhood and adolescence: From behavioral comparisons to psychological constructs to psychological comparisons. *Child Development, 52,* 129-44.

Burke, J. A., & Clark, R. A. (1982). An assessment of methodological options for investigating the development of persuasive skills across childhood. *Central States Speech Journal, 33,* 437-445.

Cantor, N., & Kilhstrom, J. F. (Eds.). (1981). *Personality, cognition, and social interaction.* Hillsdale, NJ: Lawrence Erlbaum.

Chandler, M. J. (1982). Social cognition and social structure. In F. C. Serafica (Ed.), *Social-cognitive development in context.* New York: Guilford Press.

Cline, V. B. (1964). Interpersonal perception. In B. Maher (Ed.), *Progress in experimental personality research I.* New York: Academic Press.

Crockett, W. H. (1977). *Impressions and attributions: Nature, organization, and implications for action.* Paper presented at the annual meeting of the American Psychological Association, Washington, DC.

Damon, W. (1981). Explaining social cognition on two fronts. In J. H. Flavell & L. Ross (Eds.), *Social cognitive development: Frontiers and possible futures.* Cambridge: Cambridge University Press.

Delia, J. G., O'Keefe, B. J., & O'Keefe, D. J. (1982). The constructivist approach to communication. In F.E.X. Dance (Ed.), *Human communication theory: Comparative essays.* New York: Harper & Row.

Forgas, J. P. (Ed.). (1981). *Social cognition: Perspectives on everyday understanding.* London: Academic Press.

Harvey, J. H. (Ed.). (1981). *Cognition, social behavior, and the environment.* Hillsdale, NJ: Lawrence Erlbaum.

Hastorf, A. H., & Isen, A. M. (Eds.). (1982). *Cognitive social psychology.* New York: Elsevier.

Heider, F. (1958). *The psychology of interpersonal relations.* New York: John Wiley.

Higgins, E. T., Herman, C. P., & Zanna, M. P. (Eds.). (1981). *Social cognition: The Ontario symposium.* Hillsdale, NJ: Lawrence Erlbaum.

Higgins, E. T., McCann, C. D., & Fondacaro, R. (1982). The "communication game": Goal-directed encoding and cognitive consequences. *Social Cognition, 1,* 21-37.

Higgins, E. T., Rhule, D. N., & Hartup, W. W. (Eds.). (1982). *Social cognition and social behavior: Developmental issues.* New York: Cambridge University Press.

Ickes, W., & Knowles, E. S. (Eds.). (1982). *Personality, roles, and social behavior.* New York: Springer-Verlag.

O'Keefe, B. J., & Delia, J. G. (1982). Impression formation and message production. In M. E. Roloff & C. R. Berger (Eds.), *Social cognition and communication.* Beverly Hills, CA: Sage.

O'Keefe, B. J., & Delia, J. G. (in press). Psychological and interactional dimensions of communicative development. In H. Giles & R. St. Clair (Eds.), *Recent advances in language, communication, and social psychology.* London: Erlbaum.

O'Keefe, D. J., & Sypher, H. E. (1981). Cognitive complexity measures and the relationship of cognitive complexity to communication: A critical review. *Human Communication Research, 8,* 72-92.

Roloff, M. E., & Berger, C. R. (Eds.). (1982). *Social cognition and communication.* Beverly Hills, CA: Sage.

Schutz, A. (1967). *The phenomenology of the social world* (G. Walsh & F. Lehnert, trans.). Evanston, IL: Northwestern University Press.

Schutz, A., & Luckmann, T. (1973). *The structure of the life-world* (R. M. Zaner & H. T. Engelhardt, Jr., trans.). Evanston, IL: Northwestern Univeristy Press.

Serafica, F. C. (Ed.). (1982). *Social-cognitive development in context.* New York: Guilford Press.

Tagiuri, R., & Petrullo, L. (Eds.). (1959). *Person perception and interpersonal behavior.* Stanford, CA: Stanford Univeristy Press.

About the Authors

James L. Applegate is Associate Professor and Chairperson of the Department of Communication at the University of Kentucky. He received his Ph.D. from the University of Illinois at Urbana-Champaign. His research on social cognitive and interpersonal communication processes has been published in such journals as *Human Communication Research, Communication Monographs,* and *Educational and Psychological Measurement,* as well as in the *Communication Yearbook* and numerous other books. He has been invited to present his research at over twenty national and international conventions and has conducted numerous seminars on communication practices for public and private institutions.

Mark A. Barnett is Associate Professor of Psychology at Kansas State University. He received his Ph.D. from Northwestern University. His research interests include the development and expression of prosocial behavior, children's play and make-believe, and sex-role stereotyping.

Diane Borwick received her Ph.D. in psychology at the University of Waterloo. Her research interests include parent-child communication and reading. She is currently employed as a school psychologist in Toronto, Canada.

Brant R. Burleson (Ph.D., University of Illinois at Urbana-Champaign) is Assistant Professor in the Department of Communication, Purdue University. His research interests include developmental and individual differences in functional communication skills and the social antecedents and consequences of individual differences in these skills. His articles have appeared in several communication and child development journals, including *Human Communication Research, Communication Yearbook, Communication Monographs, Quarterly Journal of Speech, Child Development, Child Study Journal, Journal of the American Forensic Association,* and *Western Journal of Speech Communication.*

Jesse G. Delia is Professor and Head of the Department of Speech Communication and a Research Professor in the Institute of Communications Research at the University of Illinois at Urbana-Champaign. He has contributed to the development of the constructivist approach to communication and to its applications to problems in social cognition, communication development, and interaction processes.

Steve Duck is Senior Lecturer in Psychology at the University of Lancaster, England, and is currently Editor of the *Journal of Social and Personal Relationships.* After receiving his B.A. and M.A. from Oxford University, he gained his Ph.D. from the University of Sheffield. He has published extensively in the areas of personal relationships, personal construct theory, the history and development of social psychology, and the dynamics of television production techniques. He edited the five-volume *Personal Relationships* series for Academic Press (the first three of them jointly with Robin Gilmour) and is currently editor (with Dan Perlman) of the Sage Series in Personal Relationships.

Nancy Eisenberg is Associate Professor of Psychology at Arizona State University. She is coauthor (with Paul Mussen) of *Roots of Caring, Sharing, and Helping: The Development of Prosocial Behavior,* and editor of *The Development of Prosocial Behavior.* Her primary research interests are in social development, including moral development, empathy, and sex-role development.

David Forbes is Director of the Peer Interaction Project at the Harvard Graduate School of Education. He is the editor of *New Direc-*

tions in Child Development: Children's Planning Strategies (Jossey-Bass, 1982). His current research interests include the development of social reasoning through peer fantasy play.

Sharon D. Herzberger is Associate Professor of Psychology at Trinity College in Hartford, Connecticut. She received her Ph.D. from the University of Illinois. She has combined her interest in the causes and consequences of child abuse with a more general interest in social cognition. Her recent research includes the perception of abused and nonabused children toward parents (*Journal of Consulting and Clinical Psychology*, 1981), the cyclical hypothesis of child abuse (in Finkelhor et al.'s *The Dark Side of Families*, 1983) and the transition to parenthood (in Friedman and DiMatteo's *Interpersonal Relations in Health Care*).

E. Tory Higgins is Professor of Psychology at New York University and a leading scholar in the area of social cognition and communication. He has edited the first three volumes of the Ontario Symposium series as well as three other important social cognition volumes. He is the author of numerous articles and monographs, and has presented his research at meetings throughout Europe, Canada, and the United States.

Thomas J. Housel is Assistant Professor of Business Communication at the University of Southern California. He received his Ph.D. in communication from the University of Utah. His areas of interest are in language behavior, memory, and cognition from a schema theory perspective. He is currently conducting research on formal communication and teleconferencing with two very large organizations.

Robert Kraut (Ph.D., Yale University, 1973) is a Research Psychologist at Bell Communications Research. He has been on the faculty at the University of Pennsylvania, Cornell University, and Princeton University. His research interests include interpersonal communication, especially deception and the coordination of conversation, and the social impact of new computer and communications technologies.

Steven H. Lewis is a social psychologist completing his Ph.D. at Yale University. His research interests include attitude change, social cognition, and judgment and decision making. At Cornell University, he worked with Robert Kraut on studies of coordination and information exchange in conversation and studies of strategies and shortcomings in social inference.

David Lubin is founder and Vice President of Interactive Training Systems, Inc., in Cambridge, Massachusetts, a company actively involved in developing laser videodisc technology for training and education in the corporate setting. His former positions include Assistant Professor in Child Study, Tufts University, and Research Associate in Education and Co-Director of the Peer Interactive Project (with David Forbes), Harvard University. He received his Ph.D. from Harvard University in 1977.

C. Douglas McCann received his Ph.D. from the University of Western Ontario and is now Assistant Professor of Psychology at York University in Toronto. His primary research interests focus on impression formation, interpersonal communication, and psychopathology. He is currently examining communication in psychopathology with a specific focus on depression.

David K. Miell is Lecturer in Social Psychology and Communication at University College, Cardiff. He obtained his Ph.D. from Lancaster University, where his focus was on empathy and other aspects of relational communication, while attached to the SSRC Research Programme on Friendship Development.

Dorothy E. Miell is Lecturer in Psychology with the Open University, Milton Keynes, U.K. She worked for four years on the SSRC Research Programme on Friendship Development at Lancaster University, and her main research interests are in the study of strategic relational communication.

Sherri Oden is Assistant Professor and Coordinator of Advanced Study in Early Childhood Education in the Graduate School at

Wheelock College in Boston, Massachusetts. She received her Ph.D. from the University of Illinois at Urbana-Champaign. Her research interests include the development of children's peer relationships and social skills training with children who lack positive peer relationships. Her recent publications include research on the development of diverse types of peer relationships in childhood (coauthored with Sharon Herzberger, Peter Mangione, and Valerie Wheeler, in *Boundary Areas in Psychology: Developmental and Social*) and the applicability of social skills training research (in *Child and Youth Services,* 1983).

Barbara J. O'Keefe is Assistant Professor of Speech Communication at the University of Illinois—Urbana, where she received her Ph.D. in speech communication. Her research interests are in communication theory, social cognition and communication, and processes of social interaction. She has contributed to the development of a general constructivist approach to communication.

Kenneth H. Rubin is Professor of Psychology and Head of the Graduate Program in Developmental Psychology at the University of Waterloo. His research interests include the development of peer relationships and social skills, social cognitive development, and children's play.

Rainer Silbereisen is Professor of Educational Psychology at the Institüt für Psychologie, Techische Universität Berlin. He has authored numerous papers on social cognition and adolescence, and is coeditor of a volume on social cognition published in German entitled *Entwicklung sozialer Kognition: Modell, Theorien, Methoden, Anwendung.*

Howard E. Sypher is Associate Professor of Communication at the University of Kentucky. He received his Ph.D. from the University of Michigan. His research interests include individual differences in communication and social cognition and communication. His recent publications have appeared in *Human Communication Research, Communication Research and Personality,* and *Social Psychology Bulletin.*

Valerie A. Wheeler is a Post-Doctoral Research Fellow and Visiting Assistant Professor in the Department of Psychology at the University of Illinois. She received her Ph.D. from the University of Rochester. Her research interests include children's perceived social competence and the process of conflict resolution in children's friendships. She recently published research on a scale developed with Gary Ladd, the Children's Self-Efficacy for Peer Interaction Scale (*Developmental Psychology*, 1982).